Safety Made Easy
3rd edition

*A Checklist Approach to
OSHA Compliance*

John R. Grubbs and Sean M. Nelson

Government Institutes
An imprint of
The Scarecrow Press, Inc.
Lanham, Maryland • Toronto • Plymouth, UK
2006

Published in the United States of America
by Government Institutes, an imprint of The Scarecrow Press, Inc.
A wholly owned subsidiary of
The Rowman & Littlefield Publishing Group, Inc.
4501 Forbes Boulevard, Suite 200
Lanham, Maryland 20706
http://www.govinstpress.com/

Estover Road
Plymouth PL6 7PY
United Kingdom

British Library Cataloguing in Publication Information Available

Library of Congress Cataloging-in-Publication Data

Grubbs, John R.
    Safety made easy: a checklist approach to OSHA compliance / John R. Grubbs and Sean M. Nelson—3rd ed.
        p.; cm.
    Includes bibliographical references and index.
    ISBN-13: 978–0–86587–162–5 (alk. paper)
    ISBN-10: 0–86587–162–0 (alk. paper)
    1. Industrial safety—Law and legislation—United States. 2. Industrial hygiene—Law and legislation—United States.
I. Nelson, Sean M. II. Title.
    [DNLM: 1. United States. Occupational Safety and Health Administration. 2. Occupational Health—legislation &
jurisprudence—United States. 3. Guideline Adherence—United States. WA 33 AA1 G885s 2006]
    KF3570.Z9D38 2006
    344.7304'65—dc22
                                                                                                2006036750

⊖™ The paper used in this publication meets the minimum requirements of
American National Standard for Information Sciences—Permanence of
Paper for Printed Library Materials, ANSI/NISO Z39.48-1992.
Manufactured in the United States of America.

For

Parker, Cameron, and Amie

Thanks for the Support and Understanding! JG

Traci, Gavin, and Ken

A wonderful wife, a precious son, and my mentor. SN

Summary Contents

Contents

Preface

Safety Made Easy, the third edition, continues to provide a product that demonstrates OSHA compliance does not have to be a *nightmare* for business professionals. We, the authors, hope to continue that tradition by providing readers with an updated book that improves compliance and more importantly gives the reader a true understanding of what it takes to provide a safe and healthy workplace. As a tool, *Safety Made Easy* provides readers with the confidence to approach any workplace and immediately reduce injuries, thus enhancing worker morale and improving productivity.

OSHA compliance is very challenging to general industry. 29 CFR 1900-1910.END is a 1,500 page two-volume set of regulations in legal jargon. Deciphering complicated regulations that reference other regulations can frustrate even the most seasoned professional. *Safety Made Easy* will assist you by providing proven interpretations of these regulations in an easy-to-read format. Also, as an added benefit, we have included proven safety recommendations that are drawn on our own experience in the field of safety.

This book is intended to provide a simple alternative that will provide the reader with an easy-to-read format that minimizes time and effort. Topics are alphabetically summarized into easy-to-understand checklists that give the reader clear direction to improving compliance. *Safety Made Easy* does not cover every topic in the Code of Regulations (CFRs), nor is it intended to replace them. It provides a simple version to aid in understanding regulations as well as achieving compliance.

This new edition of *Safety Made Easy* has been entirely updated to help the reader better keep pace with the constantly changing regulatory scene. The biggest change to the third edition is that we have provided "extra" information to the reader that we offer as recommendations to improve your safety program and ultimately attain compliance. These recommendations can be found in several sections throughout the text and are identified as Best Practices (BP), replacing the bullet point with **BP**. Some subjects such as the Ergonomics section and the new Behavior-Based Safety Section are Best Practice (BP) Sections to help employers take their programs to the next level.

As you begin or continue your compliance journey with *Safety Made Easy*, we encourage you to let us know how you are using the book and what we can do to make it the best safety tool possible. If it works for you, recommend it to a friend or colleague so that their lives might be made a little *easier* as well. Above all, BE SAFE!

THE AUTHORS

About the Authors

John Grubbs, MBA, RPIH, CSIT, is the principal consultant and owner of GCI, a full-service training and consulting firm in Longview, Texas. Specializations include safety consulting, behavior-based safety implementation, and safety leadership training for supervisors, managers, and executives. Clients include healthcare, transportation, manufacturing, education, and service organizations. You can learn more about GCI at www.gci4training.com.

John has over 13 years of leadership experience, published several books and articles, and works with leaders at all levels to improve the performance of many well-known companies internationally. He holds degrees in Occupational Safety and Health, Industrial Technology, and a Master of Business Administration. He is a Registered Professional Industrial Hygienist and a Certified Senior Industrial Technologist. John is a dynamic and energetic speaker as well as a popular trainer and business coach. Current memberships include the American Society of Safety Engineers, American Industrial Hygiene Association, National Association of Industrial Technology and the American College of Healthcare Executives.

Sean Nelson, CIT, RPIH, is a Technical Consultant for the Liberty Mutual Insurance Group, National Market Loss Prevention team responsible to a large dedicated client for the North Texas District of the Southwest Region. His primary duties as a consultant consist of behavior-based safety activities, compliance assessments, and training.

Sean's experience includes multiple years in the airline industry, manufacturing, and consulting. His biggest successes include leading a facility team toward achieving OSHA VPP (Voluntary Protection Program), developing safety policy for multiple companies, and managing safety and environmental programs at the site and corporate levels. Sean holds degrees in Industrial Technology and Occupational Safety and Health Technology. He is a Certified Industrial Technologist and a Registered Professional Industrial Hygienist and is a member of the American Society of Safety Engineers (ASSE), the National Association of Industrial Technologists (NAIT), the National Fire Protection Association (NFPA), and the Association of Professional Industrial Hygienists (APIH).

Section 1
Accident Prevention Signs and Tags

OVERVIEW

This section defines the basic requirements for accident prevention signs and tags to indicate or describe as best possible, the specific hazards in the area or within the worksite. Included in this section are specifics for the design, application, and use of signs or symbols. The intent is to raise awareness of these hazards. Failure to designate these areas may lead to accidental injury to workers or the public, or both, or to property damage.

REGULATORY COMPLIANCE

☐ §1910.145 Specifications for Accident Prevention Signs and Tags

CLASSIFICATION OF SIGNS ACCORDING TO USE

☐ Danger signs

- o Used to warn of specific dangers and radiation hazards. There shall be no variation to the type of design for these (must be consistent throughout, see "Sign Design" Subsection).

- o Instruct employees that danger signs indicate immediate danger.

☐ Caution signs

- o Warn against potential hazards or unsafe practices.

- o Instruct employees that caution signs indicate a possible hazard.

☐ Safety Instruction Signs

- o Used for general instruction and suggestions relative to safety measures.

SIGN DESIGN

☐ Design features

- o Round or blunt corners.

- o Free from sharp edges, burrs, splinters, or other sharp projections.

- o Bolts or fastening devices located in a way that will not constitute a hazard.

- o Wording must be easily read.

- o Contains enough information to be easily understood.

- o Color of background contrasting the lettering.

o   Colors shall be those of opaque glossy samples specified in Table 1 of Fundamental Specification of Safety Colors for CIE Standard Source "C", ANSI Z53.1-1967.

☐ Danger Signs

   o   White background.

   o   The word "Danger" in white letters within a black panel with red oval insert.

   o   Any letters against the white background shall be black

Black Panel with Red Insert and White Letters

White Background

Black Letters

☐ Caution Signs

   o   Yellow background.

   o   Panel black with yellow letters.

   o   Any letters against the yellow background shall be black.

Black Panel with Yellow Letters

Yellow Background

Black Letters

☐ Safety Instruction Signs

   o   White background

   o   Panel green with white letters

   o   Any letters used against white background shall be black

   o   Colors shall be opaque glossy

☐ Slow-moving Vehicle Emblem

   o   Fluorescent yellow-orange triangle.

   o   Dark red reflective border.

   o   Used only on vehicles that move slowly (25 m.p.h. or less) on public roads.

SIGN WORDING

- [ ] Easily read and concise.

- [ ] Contains sufficient information to be easily understood.

- [ ] Should make positive, rather than negative suggestion.

- [ ] Should be accurate in fact.

- [ ] Biological hazard signs shall be used to signify the actual or potential presence of a biohazard and to identify equipment, containers, rooms, materials, experimental animals, or combinations thereof, which contain or are contaminated with biological hazards or infectious substances.

BP: It may be necessary to incorporate signs in the workplace that includes a foreign language relevant to the region.

ACCIDENT PREVENTION TAGS

- [ ] Tags contain a single word and a major message that indicates the specific hazardous condition or instruction to the employee ("Danger," "Caution," "Biological Hazard," "BIOHAZARD," or the biological hazard symbol).

- [ ] Single word readable at a minimum of 5 feet or greater as warranted by the hazard.

- [ ] Major message presented in pictographs, written text, or both.

- [ ] Single word and major message shall be understandable to all employees exposed to the hazard.

- [ ] Inform employees of the meaning of various tags used throughout the workplace and what special precautions are necessary.

- [ ] Affixed as close as possible to the hazard.

- [ ] "Danger" tags – major hazard situations where immediate hazard presents threat of death or serious injury to employees.

- [ ] "Caution" tags – minor hazard situations, non-immediate threat or potential hazard or where an unsafe practice presents less of a threat of employee injury.

- [ ] "Warning" tags – hazard between "Danger" and "Caution" level.

- [ ] "Biological Hazard" tags – actual or potential presence of a biological hazard, and to identify equipment, containers, rooms, experimental animals, or a combination of these that contain or are contaminated with hazardous biological debris or fluids.

- [ ] Other tags – other tags may be used to provide additional information to the employee provided that they do not detract from the impact or visibility of the single word or major message of any required tag.

ANSI Z-535.2—1998 STANDARD FOR ENVIRONMENTAL AND FACILITY SAFETY SIGNS

☐ Environmental and facility safety signs shall have a consistent appearance. (Use of previous standard from 1991 is still acceptable)

☐ **"Danger"** or **"Warning"** headers or signal panels shall incorporate the safety alert symbol to indicate a potential for personal injury.

☐ Safety alert symbol is optional in caution headers and shall not be used to identify hazards that will result only in property damage.

☐ Message text shall be left-margin aligned and incorporate mixed cases for increased readability. "Sans Serif" fonts (e.g. T, but not **T**) should be used.

☐ Symbols should be incorporated into the message text to enhance communication.

☐ Signs shall incorporate a banner in an appropriate color and message text shall appear on a white background.

COLOR IDENTIFICATION/DEFINITION

☐ Red

- o Fire protection equipment and apparatus
- o Danger
- o Stop
- o Flammable liquids
- o Flammable gases

☐ Yellow

- o Caution
- o Oxidizing agents
- o Striking
- o Falling or tripping hazard
- o Being caught between

☐ Orange

- o Warning
- o Caution
- o Danger

☐ Fluorescent orange or red orange

- o Materials
- o Containers
- o Equipment
- o Areas or rooms

☐ Green
- o Safety

---------- END OF SECTION ----------

Section 2
Behavior-Based Safety (BBS)

OVERVIEW

This section provides information regarding behavior-based safety. Behavior-based safety (BBS) is a common approach to gain employee participation and buy-in with a safety program. The ultimate goal of a BBS system is to develop a culture where employees exhibit safe behaviors as habit or routine of everyday life. This section will describe common approaches, misconceptions, and methodologies for BBS.

REGULATORY COMPLIANCE

The information in this section is provided as a best management practice (**BP**). There are no current OSHA regulations that prescribe the use or requirements of a BBS system. In addition, BBS systems may not be appropriate for all organizations.

WHAT IS BBS?

☐ A method to develop a true safety culture within an organization that is ready for change.

☐ Focus is on behaviors rather than conditions as primary causes of accidents.

☐ Encourages employee participation in the safety program by conducting safety observations to identify safe and at-risk behaviors.

☐ Avoids blame by placing emphasis on positives through anonymous observation cards.

☐ Uses psychology to implement individual change by tapping into employee perceptions.

☐ Introduces values as a foundation for change that promotes easy decision making for the organization.

☐ Utilizes hard data to track and trend organizational change through statistics that are leading indicators rather than trailing indicators.

COMMON APPROACHES

☐ Trained observers are encouraged to conduct and document observations of other employees in the workplace.

☐ Success is gained by the conversations taking place between employees and the discussion of safe and at-risk behaviors on the job. In some cases, employees are unaware that their actions are of an at-risk nature.

☐ Provide positive feedback first for well-done activities or wanted behaviors (e.g. "Jack, I've noticed you do a great job of bending at the knees to pick up heavy boxes."). For unwanted or at-risk behaviors, provide constructive feedback last (e.g. "...however, Jack, you tend to twist a bit rather than pivoting when placing the boxes on the conveyor."). Constructive feedback for unwanted or at-risk behaviors can also be provided with a positive twist (e.g. "Jack, I observed you lifting correctly by bending at the eight out ten times."). The employee hears the positive feedback, but realizes that there is opportunity for improvement.

☐ Observation cards (anonymous) are collected to identify trends and promote emphasis on issues that are most common in the organization.

  o An organization that has more "safe" observations regarding the use of personal protective equipment need not focus on this topic for training.

  o An organization that has more "at-risk" observations for posture and lifting can place emphasis in this area for training.

☐ Statistics are tracked to provide feedback to the organization regarding the quality and quantity of observations.

☐ Front-line employee-driven approaches are most successful as employees take ownership for the process and it resists the temptation of becoming another management fad or program that fades away with time.

MISCONCEPTIONS

☐ Easy to implement – It will certainly take time to change your culture. An organization can expect 12-18 months to see an organizational impact.

☐ Quick fix – Organizations can and should expect activity peaks and valleys. Leaders at all levels must encourage and continually champion the process for success and sustainability.

☐ One shot – The process needs continual follow-up to maintain effort for success. Retraining and communication is needed to bring new participants into the process.

☐ Can be forced – Observation quality diminishes with quotas. Organization should only encourage participation.

CONCLUSION

☐ An organization must be assessed thoroughly at all levels before implementing a BBS process.

☐ A strong basic safety program with a disciplined workforce must be in place before a BBS process is implemented.

☐ Copious amounts of training to promote buy-in at all levels is required to avoid the process gaining a negative or "snitch sheet" perspective from employees.

☐ Strong follow-up metrics are required to ensure long-term viability in the organization.

---------- END OF SECTION ----------

Section 3
Bloodborne Pathogens (Exposure Control)

OVERVIEW

This section addresses occupational exposure to blood and other potentially infectious substances. This section applies only to personnel who are (reasonably) expected to come into contact with such substance such as doctors, nurses, health care providers, nursing home attendants, or designated first aid responders trained to respond to incidents/accidents. Most facilities will only have personnel who fit into the last category.

Clean up measures for incidental releases of blood or other potentially infectious substances are identified as "Universal Precautions." Guidelines for universal precautions can be found in a subsection by that name later in this section.

REGULATORY COMPLIANCE

☐ §1910.1030 Bloodborne Pathogens

TERMS/DEFINITIONS

☐ **_Blood_** - human blood, human blood components, and products made from human blood.

☐ **_Bloodborne Pathogens_** - pathogenic microorganisms that are present in human blood and can cause disease in humans. These pathogens include, but are not limited to, hepatitis B virus (HBV) and human immunodeficiency virus (HIV). *Custodial duties that include removal of feminine hygiene products do not fall under the scope of the BBP Standard and will not require (on its own) a written exposure control plan (per OSHA interpretation letter).*

☐ **_Contaminated_** - the presence or the reasonably anticipated presence of blood or other potentially infectious materials on an item or surface.

☐ **_Decontamination_** - the use of physical or chemical means to remove, inactivate, or destroy bloodborne pathogens on a surface or item to the point where they are no longer capable of transmitting infectious particles, and the surface or item is rendered safe for handling, use, or disposal.

☐ **_Exposure Incident_** - a specific eye, mouth, other mucous membrane, non-intact skin, or parenteral contact with blood or other potentially infectious materials that result from the performance of an employee's duties.

☐ **_Hand Washing Facilities_** - a facility providing an adequate supply of running potable water, soap, and single-use towels or hot air drying machines.

☐ **_Licensed Health Care Professional_** - a person whose legally permitted scope of practice allows him or her to independently perform the activities required by the Standard for Hepatitis B Vaccination and Post-exposure Evaluation and Follow-up.

☐ *HBV* - hepatitis B virus.

☐ *HIV* - human immunodeficiency virus.

☐ *Occupational Exposure* - reasonably anticipated skin, eye, mucous membrane, or parenteral contact with blood or other potentially infectious materials that may result from the performance of an employee's duties.

☐ *Other Potentially Infectious Materials* - (1) The following human body fluids: semen, vaginal secretions, cerebrospinal fluid, synovial fluid, pleural fluid, pericardial fluid, peritoneal fluid, amniotic fluid, saliva in dental procedures, any body fluid that is visibly contaminated with blood, and all body fluids in situations where it is difficult or impossible to differentiate between body fluids; (2) Any unfixed tissue or organ (other than intact skin) from a human (living or dead); and (3) HIV-containing cell or tissue cultures, organ cultures, and HIV- or HBV-containing culture medium or other solutions; and blood, organs, or other tissues from experimental animals infected with HIV or HBV.

☐ *Parenteral* - piercing mucous membranes or the skin barrier through such events as needle sticks, human bites, cuts, and abrasions.

☐ *Personal Protective Equipment (PPE)* - specialized clothing or equipment (safety shoes, glasses, aprons, etc.) worn by an employee for protection against a hazard. General work clothes (e.g., uniforms, pants, shirts or blouses), not intended to function as protection against a hazard, are not considered to be personal protective equipment.

☐ *Regulated Waste* - liquid or semi-liquid blood or other potentially infectious materials; contaminated items that would release blood or other potentially infectious materials in a liquid or semi-liquid state if compressed; items that are caked with dried blood or other potentially infectious materials and are capable of releasing these materials during handling; contaminated sharps; and pathological and microbiological wastes containing blood or other potentially infectious materials.

☐ *Universal Precautions* - an approach to infection control. According to the concept of Universal Precautions, all human blood and certain human body fluids are treated as if known to be infectious for HIV, HBV, and other bloodborne pathogens.

☐ *UPK* - Universal Precaution Kit.

☐ *Work Practice Controls* - controls that reduce the likelihood of exposure by altering the manner in which a task is performed (e.g., prohibiting recapping of needles by a two-handed technique).

TRAINING / QUALIFICATIONS

☐ The training program shall contain at a minimum the following elements:

- o The location of an accessible copy of the regulatory text of the 1910.1030 standard and an explanation of its contents;

- o A general explanation of the epidemiology and symptoms of bloodborne diseases;

- o An explanation of the modes of transmission of bloodborne pathogens;

- o An explanation of the employer's exposure control plan and the means by which the employee can obtain a copy of the written plan;

- o An explanation of the appropriate methods for recognizing tasks and other activities that may involve exposure to blood and other potentially infectious materials.

☐ Training will be provided prior to performing duties applicable to this section and at least biannually (every two years) thereafter.

UNIVERSAL PRECAUTIONS

A bloodborne pathogen is a microorganism in the blood or bodily fluids (spinal fluid, etc.) that can cause disease. Two microorganisms of particular concern are the Human Immunodeficiency Virus (HIV) and the Hepatitis B Virus (HBV). While chances of coming in contact with these viruses in most jobs may not be anticipated under normal circumstances, it is important to minimize the chances of exposure. For initial handling, the "Universal Precaution" is to assume all blood and bodily fluids are contaminated and to avoid direct contact. Utilizing Universal Precautions and a Universal Precaution Kit (UPK) will decrease the chances of contact with blood or bodily fluids.

☐ Universal precautions shall be observed to prevent contact with blood or other potentially infectious materials. Under circumstances in which differentiation between body fluid types is difficult or impossible, all body fluids shall be considered potentially infectious materials.

☐ When there is occupational exposure (potential for contact), the employer shall provide, at no cost to the employee, appropriate personal protective equipment (PPE) such as, but not limited to, gloves, face shields or masks, eye protection, mouthpieces, pocket masks, and/or other ventilation devices. PPE will be considered "appropriate" only if it does not permit blood or other potentially infectious materials to pass through to or reach the employee's work clothes, street clothes, undergarments, skin, eyes, mouth, or other mucous membranes under normal conditions of use and for the duration of time which the protective equipment will be used.

☐ When provision of hand washing facilities is not feasible, the employer shall provide either an appropriate antiseptic hand cleanser in conjunction with clean cloth/paper towels or antiseptic towelettes. When antiseptic hand cleansers or towelettes are used, hands shall be washed with soap and running water as soon as feasible.

☐ If an exposure or spill of blood or other potentially infectious substance occurs, use universal precautions to handle, clean, and/or dispose of contaminated surfaces or materials. See "Universal Precautions" below for specific instruction.

☐ Universal Precaution Kit (UPK)
A UPK should (at a minimum) contain the following components:

o Safety shield

o Identification tag (not required)

o Medical gloves

o Red bag

o Red Z® powder and pouch

o Protective apron

o Scoop and scraper

o Disinfectant solution with wipe

o Anti-microbial towelette

BP: Recommended universal precaution procedures:
For an incident involving the release of bodily fluids, notify your supervisor as soon as practical and perform clean up by following the instructions below. Instructions are also located on the UPK.

- o Put on the disposable apron, safety shield, and gloves. Place the safety shield over the face so that the shield covers the eyes and the mask covers the mouth.

- o Sprinkle Red-Z® powder over spilled area and allow the spill to solidify to a dry gel.

- o Remove the gelled material using the scoop and scraper and carefully place in a regular trash bag. This bag will serve as an inner liner or secondary containment.

- o Clean away remaining solids and disinfect any affected surface area with the disinfectant solution and dry wipe (follow instructions on the disinfectant/deodorant before use).

- o Seal and dispose using the steps below.

- o Promptly wash hands thoroughly with soap and water or anti-microbial towelette included in the kit.

> **NOTE:** Do not use the Red-Z® powder on areas that may come into contact with food items (tables used for eating, etc.).

BP: Recommended procedures for disposal of UPK items:
Disposal of the kit is as follows:

- o If the substance does not contain blood, the red bag is unnecessary. Use a regular (non-red) trash bag. Discard the bag into normal trash. Dispose of all other unused kit items.

- o If the substance contains visible blood (or if unsure), use the red bag from the kit to store all the used articles, seal the bag, and secure it. Contact your Safety & Health representative as soon as practical for disposal.

- o Restock replacement UPKs as necessary.

- o Complete an incident report.

POST-EXPOSURE (DIRECT CONTACT) EVALUATIONS AND FOLLOW-UP

The information below is a combination of requirements and recommendations to help better comply with the complications of this Standard. To see the specifics of these requirements, please refer to the 29CFR 1910.1030 Standard.

BP: If an exposure occurs (see definition of "Exposure Incident"), first thoroughly clean the area with soap and water for at least 30 seconds. Then, take these actions:

- o The exposed individual should notify his/her manager/supervisor.

- o The manager/supervisor should coordinate with the Safety & Health representative to complete a bloodborne pathogens report.

- o A licensed health care professional will perform an evaluation and medical follow-up at no cost to the employee. If an occupational health nurse is not available, the individual may be sent to local hospital or clinic by his/her manager/supervisor for immediate attention.

- o A Post-Exposure Testing Consent/Refusal Form (many occupational health clinics can provide or one can be produced by the employer) must be signed indicating consent or declination for blood collection and testing for Hepatitis B and/or HIV. Employees have the right to refuse testing. If the exposed employee gives consent for blood collection but not for HIV testing, the blood is kept for 90 days, during which time the employee can choose to have the sample tested at no cost to the employee.

BP: If all tests are negative, the employee is considered not at risk. However, further evaluation may be recommended by the evaluating health care professional. If so, the following applies:

- o The exposed employee shall be offered treatment and further evaluation if the evaluating health care professional believes it is needed. This care shall be provided at no cost to the employee.

o   A written opinion by the evaluating health care professional stating that the employee has been informed of the results of the evaluation and about any exposure-related conditions that will need further evaluation and treatment. This information will be included in the employee's medical record. This record is maintained by the employer in a restricted access location. All medical records shall be kept strictly confidential.

HEPATITIS B POST-EXPOSURE IMMUNIZATION

☐   If an employee is involved in an incident with possible HBV exposure, a vaccine shall be offered at no cost to the employee.

o   The Hepatitis B Immunization Consent/Refusal Form (many occupational health clinics can provide or one can be produced by the employer) must be signed.

o   The employer shall ensure that the signed form is included in the medical record maintained by the employer and kept confidential.

o   The post-exposure immunization shall be arranged by the employer and provided free of charge to the employee.

----------END OF SECTION----------

Section 4
Chains, Slings & Cables

OVERVIEW

This section provides the requirements for the safe use of alloy steel chain, wire rope, metal mesh, and natural or synthetic fiber rope slings.

REGULATORY COMPLIANCE

☐ §1910.184 Slings

SAFE OPERATING PRACTICES

☐ Do not use defective slings.

☐ Do not shorten a sling with knots, bolts, or other makeshift devices.

☐ Do not allow slings to become kinked.

☐ Never use a sling to lift more than the sling's rated capacity.

☐ Make sure slings use to form a basket hitch keep the load balanced to prevent slippage.

☐ Slings should be securely attached to their loads.

☐ Slings shall be padded or protected from sharp edges.

☐ Keep suspended loads clear from all obstructions.

☐ Employees should be kept clear of lifted or suspended loads.

☐ Do not place hands and fingers between the sling and the load.

☐ Employees should never shock load while lifting.

☐ Never pull a sling from under the load while the load is resting on the sling.

INSPECTIONS

☐ Each day before being used, a competent person shall inspect the sling and all fastening components for damage or defects.

☐ Additional inspections may be required during use.

☐ Damaged or defective slings shall be immediately removed from service.

ALLOY STEEL CHAIN SLINGS

☐ Alloy steel slings shall have a permanently affixed durable identification stating size, grade, rated capacity, and reach.

☐ All attachments must have a rated capacity equal to the sling.

☐ Makeshift attachments such as bolts cannot be used.

☐ Alloy steel slings must be inspected and documented at least once every 12 months.

☐ If heated above 1000° F, the sling must be removed from service.

BP: If heated above 600° F, the sling should be removed from service. Additionally it is not recommended that alloy steel slings be reconditioned or repaired.

☐ Slings with cracked or deformed master links or coupling links shall be removed from service.

☐ Slings with damaged hooks or hooks open more than 15 percent must be removed from service.

WIRE ROPE SLINGS

☐ Consult the manufacturer for specific damage to look for during pre-use inspection of your type of cable.

☐ Make sure attachments meet the same load standards as the cable.

☐ Fiber-core wire ropes shall be removed from service if they are exposed to temperatures in excess of 200° F.

☐ Non-fiber core wire ropes shall be used only at temperatures below 400° F and above 60° F. Recommendations by the manufacturer regarding use at that temperature shall be followed.

METAL MESH SLINGS

☐ Handles shall have a rated capacity at least equal to the metal fabric.

☐ The fabric and handles shall be joined so that the rated capacity of the sling is not reduced, the load is easily distributed, and sharp edges will not damage the fabric.

☐ Metal mesh slings not impregnated with elastomers may be used in a temperature range from minus 20° F to 550° F.

☐ Metal mesh slings impregnated with polyvinyl chloride or neoprene may be used in a temperature range of 0° F to 200° F.

BP: Always follow manufacturer's recommendations when using slings.

NATURAL OR SYNTHETIC FIBER ROPE SLINGS

☐ These slings may be used only in temperatures of above minus 20° F to 180° F unless they are wet and frozen.

☐ Never splice these slings (only the manufacturer may alter slings).

☐ Natural or synthetic fiber slings shall be removed from service if:

    o   There is abnormal wear

    o   Powdered fibers appear between strands

    o   Fibers are broken or cut

    o   Variation in size or roundness of strands occurs

    o   Discoloration or rotting is detected

    o   Distortion of hardware is detected

SYNTHETIC WEBSLINGS

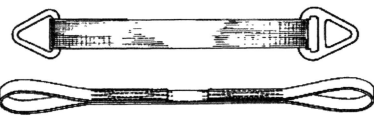

☐ Make sure all webbing is uniform in thickness.

☐ Polyester and nylon webs shall not be used where fumes, vapors, sprays, mists, liquids of acids, phenolics, or caustics are present.

☐ Remove from service when the following conditions are present:

- o Acid or caustic burns
- o Melting or charring of any part of the sling
- o Snags, punctures, tears, or cuts
- o Broken or worn stitches
- o Distortion of any fitting

RECORDKEEPING

☐ Inspection records must be maintained.

☐ Proof Test Certification from manufacturer or testing agent records must also be maintained.

----------END OF SECTION----------

Section 5
Compressed Gases

OVERVIEW

The purpose of this section is to describe guidelines, rules, and regulations set forth by OSHA that pertain to the inspection, handling, storage, marking, and labeling of compressed gas cylinders. The hazards associated with compressed gas cylinders can be controlled and minimized if the recommendations of this section are followed. NOTE: This section does not cover industrial piping systems. In addition, this section will not cover the manufacture of or the design specifications for compressed gas cylinders.

REGULATORY COMPLIANCE

☐ §1910.101 Compressed Gases (General Requirements)

☐ §1910.102 Acetylene

☐ §1910.103 Hydrogen

☐ §1910.104 Oxygen

☐ §1910.105 Nitrous Oxide

☐ §1910.252 General Requirements (for Welding, Cutting, and Brazing)

☐ §1910.253 Oxygen-fuel Gas Welding and Cutting

☐ National Fire Protection Association (NFPA) 51B & 55

GENERAL INFORMATION

☐ Instructions - For equipment that requires operation by employees, clear operating instructions shall be maintained and readily available.

☐ Maintenance/Storage
  o The equipment and functioning of each system shall be maintained in a safe operating condition in accordance with the requirements of this section.
  o Storage containers, piping, valves, regulating equipment, and other accessories shall be protected against physical damage and tampering.
  o Mobile equipment used as permanent equipment shall be adequately secured to prevent movement.

o   Mobile equipment shall be bonded to dissipate static electricity.

☐ Visual and other inspections shall be conducted as prescribed by the appropriate regulatory agency, depending on the purpose and use of the cylinder in question. (e.g., for shipping and transportation, the Hazardous Materials Regulations [49CFR] should be referenced. For Industrial use and storage, the Compressed Gas Association pamphlets [C-6-1968 and C-8-1962] should be referenced.)

☐ Markings – Compressed gas cylinders shall be legibly marked for the purpose of identifying the gas content with either the chemical or the trade name of the gas. Such marking shall be by means of stenciling, stamping, or labeling and shall not be readily removable. Whenever practical, the marking shall be located on the shoulder of the cylinder. The numbers and markings stamped onto cylinders shall not be altered.

GENERAL PROCEDURES AND PRECAUTIONS

Mixtures of fuel gases and air or oxygen may be explosive and shall be avoided. Due to this possible hazard, the following precautions shall be observed:

☐ Mixing

o   No device permitting the mixture of air or oxygen with flammable gases prior to consumption, except at the burner or in a standard torch, shall be allowed unless approved for the purpose.

o   No person other than the gas supplier shall attempt to mix gases in a cylinder.

o   No one, except the owner of a cylinder or a person authorized by him, shall refill a cylinder.

☐ Equipment

o   Only approved apparatuses such as torches, regulators, or pressure-reducing valves shall be used.

o   No one shall tamper with safety devices in cylinders or valves.

o   Before connecting a regulator to a cylinder valve, the cylinder valve shall be opened slightly for an instant, to clear dirt and dust, and then closed. Always stand to one side of the outlet when opening the cylinder valve. A fuel gas cylinder valve shall never be opened near welding, sparks, flame, or other sources of ignition.

o   Before a regulator is removed from a cylinder valve, the cylinder valve shall be closed and the gas released from the regulator.

☐ Cylinder Valve

o   The cylinder valve shall always be opened **slowly**. Cylinder valves, except acetylene, shall be fully opened to prevent leakage around the stem.

o   If cylinders are found to have leaky valves that cannot be stopped by closing the valve, the cylinders shall be taken outdoors, away from sources of ignition, and slowly emptied.

o   A hammer or wrench shall not be used to force cylinder valves. If valves cannot be operated by hand, they shall be considered defective and taken out of service.

o   Complete removal of the stem from a diaphragm-type cylinder valve shall be avoided.

☐ Handling

o   Cylinders shall not be dropped, struck, or permitted to strike each other.

o   Cylinders shall be handled carefully. Rough handling, knocks, or falls may damage the cylinder, valve, or safety devices and cause leakage.

- o Cylinder valves shall be closed when work is finished or before moving the cylinder.
- o Cylinders shall be kept far enough away from welding or cutting operations so sparks, hot slag, or flame will not reach them.
- o Cylinders, full or empty, shall never be used as rollers or supports.

☐ Placement/ Storage
- o Containers shall be protected against damage from falling objects or work activity in the area.
- o Cylinders, cylinder valves, couplings, regulators, hose, and apparatus shall be kept free from oily or greasy substances.
- o Cylinders shall not be placed where they might come in contact with any electrical circuit or device. Cylinders shall be kept away from radiators, piping system, layout tables, etc., that may be used for grounding electricity circuits such as for arc welding machines. Any practice such as the tapping of an electrode against a cylinder to strike an arc shall be prohibited.
- o Nothing shall be placed on top of a cylinder, when in use, that may damage the safety device or interfere with the quick closing of the valve.
- o Cylinders not having fixed hand wheels shall have keys, handles, or nonadjustable wrenches on valve stems while in service. When a special wrench is required, the wrench shall be left in place on the stem of the valve while the cylinder is in use so that the gas flow can be quickly turned off in case of an emergency. In the case of cylinders connected to a manifold or coupled cylinder, at least one such wrench shall always be available for immediate use.
- o Valve protection caps shall always be in place except when a regulator is installed.

SPECIFIC PROCEDURES AND PRECAUTIONS

☐ Acetylene – The following precautions are specific to the use of acetylene fuel gas:
- o An acetylene cylinder valve shall not be opened more than 1-1/2 turns of the spindle and preferably not more than 3/4 turn.
- o Under no condition shall acetylene be generated, piped, or utilized at a pressure in excess of 15-psi gauge pressure or 30-psi absolute pressure. Acetylene above 15-psi in 8-inch cylinders is dissolved in solvent to remain stable.
- o An acetylene cylinder shall be used in an upright position to avoid loss of cylinder solvent.

☐ Hydrogen – The following precautions are specific to the use of hydrogen gas:
- o Permanently installed containers shall be provided with substantial noncombustible supports on firm noncombustible foundations.
- o Safety relief devices shall be arranged to discharge upward and unobstructed to the open air. This requirement does not apply to DOT Specification containers having an internal volume of two (2) cubic feet or less.
- o Mobile hydrogen supply units shall be electrically bonded to the system before discharging hydrogen.
- o Systems shall not be located near flammable liquid piping or other flammable gases.
- o Hydrogen gas is highly flammable and shall be kept away from spark and/or open flames.

☐ Oxygen – The following precautions are specific to the use of oxygen gas:
- o Oxygen cylinders or apparatus shall not be handled with oily hands or gloves.

- o Cylinders, cylinder valves, couplings, regulators, hose, and apparatuses shall be kept free from oily or greasy substances.

- o Unless connected to a manifold, oxygen from a cylinder shall not be used without first attaching an oxygen regulator to the cylinder valve.

- o A jet of oxygen must never be permitted to strike an oily surface or greasy clothes or enter a fuel, oil, or other storage tank.

- o Thread-sealing compounds and/or leak-check solutions, when used, shall be suitable for use with oxygen.

TRANSFER OF CYLINDERS

☐ Lifting – Care shall be exercised to prevent dropping, which might cause damage to the cylinder, valve, or safety devices. Lifting magnets, slings of rope or chain, or any other device in which the cylinders themselves form a part of the carrier shall not be used for hoisting cylinders.

☐ Protection Cap – Unless cylinders are secured on a special truck/cart, regulators shall be removed and valve-protection caps, when provided for, shall be put in place before the cylinders are moved.

☐ Cart – Except for smaller cylinders that can be transported by hand, the transfer of cylinders from storage to point of usage shall be by the cart on which the cylinder is used or some other acceptable cart/cradle. The acceptability of a cart/cradle shall be determined by the local user prior to using it to transport cylinders. Carts/cradles are acceptable if they meet the following criteria:

- o Are mechanically sound and currently in good repair

- o Have sufficient ability to support the weight of the cylinders to be transported

- o Have sufficient means to mechanically secure the cylinders to prevent falling from the cart, or striking the cart or anything else violently

☐ Cylinders on Carts with Regulators – The following are additional concerns for cylinders that are on carts and mobile when the regulator is in place:

- o The cylinder valve and regulator shall have mechanical protection from damage.

- o The cylinder shall be held in an erect/nearly erect position unless the cart is specifically designed to hold cylinders in a horizontal position.

GENERAL STORAGE OF CYLINDERS

☐ Location and Placement

- o Cylinders shall be kept away from radiators and other sources of heat.

- o Inside of buildings, cylinders shall be stored in a well-protected, well-ventilated, dry location, at least 20 feet away from highly combustible materials such as oil, excelsior, or grease.

- o Cylinders shall be stored in assigned places away from stairs or where they may be exposed to excessive rise in temperature, physical damage, or tampering by unauthorized persons.

- o Assigned storage spaces shall be equipped with safety chains or other means so that cylinders cannot be knocked over or damaged by passing or falling objects.

- o Cylinders shall not be kept in unventilated enclosures such as lockers or cabinets.

o   Special spaces shall be allocated for the storage of full and empty cylinders with provisions for a retaining chain to secure the cylinders upright. Empty and full cylinders shall be stored separately, with empty cylinders clearly marked.

☐ Valves

o   The valves on empty cylinders shall remain closed.

o   Valve protection caps, where cylinders are designed to accept a cap, shall always be in place and hand-tightened, except when cylinders are in use or connected for use. The protective valve cap should "bottom out" securely against the shoulder of the cylinder.

☐ Marking/Posting

o   Marking the letters "MT" for empty on the cylinder with chalk is recommended. Or, a position within the storage area marked for empty cylinders could be considered.

o   Storage carts and fixed-mount storage shall have content, pressure range, and precautions posted.

o   Signs shall be conspicuously posted in storage areas reading "DANGER – NO SMOKING, MATCHES, OR OPEN FLAMES" or other equivalent wording.

⬭ DANGER
OXYGEN, NO SMOKING
NO OPEN FLAMES

☐ Safety Devices

o   To keep cylinder safety-relief devices in reliable operating condition, and to prevent damage, use caution when storing cylinders.

o   Exercise care to avoid plugging of safety relief device channels or other parts by paint or dirt accumulation that could interfere with functions of the device.

SPECIFIC STORAGE OF OXYGEN CYLINDERS

☐ Storage areas can vary from a specially constructed room just for oxygen to an outside area that provides protection from the items listed here. Any area used to store oxygen shall be exclusive to oxygen. This means there shall be separation between the oxygen and any other incompatible materials as described in this section.

☐ If oxygen is kept in a room or an enclosed area, the room/area shall have ventilation that is sufficient to prevent oxygen levels in excess of 23%.

☐ Oxygen shall not be kept near reserve stocks of fuel gas cylinders or any other highly combustible material, especially oil and grease.

☐ When oxygen cylinders are stored in the same general area as any petroleum products (oil, grease, etc.) or other compressed gas cylinders, or large stocks of ordinary combustible materials, they shall be separated from those materials by 20 feet or by a noncombustible barrier at least 5 feet high having a fire resistance rating of at least ½ hour.

> **NOTE** – Barrier construction shall meet the following minimum requirements: 2½-inch metal studs (3½-inch are preferred) spaced every 24 inches on center, with one layer of gypsum wallboard on each side.

☐ Oxygen storage areas shall be clearly placarded "OXYGEN – NO SMOKING – NO OPEN FLAMES" or placards that convey an equivalent message.

☐ Cylinders shall be stored so that they are never allowed to reach a temperature exceeding 125°F. When stored in the open, they shall be protected against direct rays of the sun and from the ground beneath to prevent rusting where applicable.

CYLINDER INSPECTION (GENERAL GUIDELINES)

☐ Visual and other inspections shall be conducted as prescribed by the appropriate regulatory agency, depending on the purpose and use of the cylinder in question. (e.g., for shipping and transportation, the Hazardous Materials Regulations [49CFR] should be referenced. For Industrial use and storage, the Compressed Gas Association pamphlets [C-6-1968 and C-8-1962] should be referenced.)

☐ Each employer shall determine if compressed gas cylinders under his/her control are in a safe condition to the extent that this can be determined by visual inspection.

BP: Inspection experience is an important factor in determining the acceptability of a given cylinder for continued service. Users lacking this experience who have doubtful cylinders should return them to the supplier or manufacturer for re-inspection.

☐ All hoses shall be maintained in a safe and good working condition.

☐ All regulators shall be inspected for external corrosion, damage, plugging of pressure-relief devices, mechanical defaults, and leakage prior to use.

AIR RECEIVERS

☐ Application – This subsection applies to compressed air receivers and other equipment used in providing and utilizing compressed air for performing operations such as cleaning, drilling, hoisting, and chipping (e.g. Shop Air, etc.).

☐ Installation and Equipment Requirements

  o Installation – Air receivers shall be installed so all drains, handholds, and manholes are easily accessible. Under no circumstances shall an air receiver be buried underground or located in an inaccessible place.

  o Drains and Traps

    ◊ A drainpipe and valve shall be installed at the lowest point of every air receiver to provide for the removal of accumulated oil and water.

    ◊ Adequate automatic traps may be installed in addition to drain valves.

    ◊ The drain valve on the air receiver shall be opened and the receiver completely drained frequently (i.e., at such intervals as to prevent the accumulation of excessive amounts of liquids in the receiver).

☐ Gauges and Valves

  o Every air receiver shall be equipped with an indicating pressure gauge (so located as to be readily visible) and with one or more spring-loaded safety valves. The total relieving capacity of these valves shall prevent pressure in the receiver from exceeding the maximum allowable working pressure of the receiver by more than 10%.

  o No valve of any type shall be placed between the receiver and its safety valve or valves.

  o Safety appliances such as safety valves, indicating devices, and controlling devices shall be constructed, located, and installed so that they cannot be readily rendered inoperative by any means, including weather damage.

---------- END OF SECTION ----------

Section 6
Confined Spaces (Permit-Required)

OVERVIEW

This section provides information regarding the requirements to enter and work in permit-required confined spaces. This section also helps employers understand the difference between a confined space and a permit-required confined space. These spaces are regulated differently for agriculture, construction, and shipbuilding. Information for how these spaces are to be managed can be found in 29 CFR 1928, 1926, and 1915, respectively.

REGULATORY COMPLIANCE

☐ §1910.146 Permit-Required Confined Spaces

TERMS AND DEFINITIONS

☐ **_Acceptable entry conditions_** – conditions in a permit space that allow employees to safely enter and perform work.

☐ **_Attendant_** – an individual stationed outside one or more permit spaces to monitor authorized entrants.

☐ **_Authorized entrant_** – an employee who is authorized by the employer to enter a permit space.

☐ **_Blanking or blinding_** – the absolute closure of a pipe, line, or duct by placing a solid plate (skillet) between two flanges to cover the bore in order to prevent flow. It is important to make sure the skillet can withstand the pressure.

☐ **_Entry supervisor_** – the trained individual who is responsible for coordination of all work activities in permit required confined spaces.

☐ **_Entry permit_** – the written document that authorizes work to be performed in a permit-required confined space. Entry permits expire at the end of one shift or whenever a space has been vacated for more than thirty minutes.

☐ **_Hot work permit_** - the employer's written authorization to perform operations such as riveting, welding, cutting, burning, and heating, i.e., operations that are capable of providing a source of ignition.

CONFINED SPACE CHARACTERISTICS

☐ In order to be a confined space, an area must have ALL of the following characteristics:

  o Large enough and so configured that an employee can enter and perform assigned work

  o Has limited or restricted means of entry or exit (e.g., tanks, vats, vessels, silos, storage bins, hoppers, vaults, pits)

  o Is not designed for continuous employee occupancy

  BP: A confined space that doesn't qualify as a permit-required confined space need not be managed by this section.

PERMIT-REQUIRED CONFINED SPACE CHARACTERISTICS

☐ In order to be considered a permit-required confined space (permit space), a confined space as defined above may have ANY of the following characteristics:

  o Contains or has potential to contain a hazardous atmosphere (example: a space that contains a natural gas valve).

  o Contains a material that has a potential for engulfing an entrant (example: materials stored in bulk such as powders or liquids).

  o Has an internal configuration that could trap or asphyxiate an entrant, such as inwardly converging walls or a floor which slopes downward and tapers to a smaller cross section (example: a silo that tapers at the bottom).

  o Contains any other recognized serious safety or health hazard.

  BP: If any area meets the definition of a confined space and a hazard is present, manage the area as a permit space.

GENERAL REQUIREMENTS

☐ Identify, evaluate, and monitor the hazards in permit-required confined spaces.

☐ Post danger signs stating "DANGER—PERMIT-REQUIRED CONFINED SPACE" around permit space to inform employees of the dangers posed by the existence of and location of the permit space.

☐ Take effective measures to prevent unauthorized entry.

☐ Re-evaluate a non-permit confined space when the atmosphere changes, or the use or configuration changes to increase the hazard to entrants. If necessary, reclassify as a permit-required confined space.

☐ Inform contractors performing work in or near a permit space of the entry requirements, precautions, and procedures, and debrief contractors of entry operations and hazards confronted or created.

☐ Ensure that the contractor meets or exceeds all requirements of the company's confined space program.

☐ Have a written program available for employees to review upon request.

WRITTEN PROGRAM ELEMENTS

☐ Identify all permit-required confined spaces on a written inventory.

☐ Review the inventory as necessary to ensure accuracy.

☐ Identify the responsibilities of an entry supervisor, attendant, and entrants.

☐ Develop entry procedures to ensure each entry is performed in a safe manner.

☐ Develop and use an entry permit system for permit-required confined space entry.

☐ Provide all equipment needed to safely enter spaces to perform work.

☐ Determine rescue methods to be used in case of an emergency in a confined space.

☐ Develop a training program for entry supervisors, attendants, and entrants.

☐ Develop a policy for coordination of work with contractors.

☐ Review the program annually.

☐ Keep records of training and expired permits as required.

INVENTORY OF PERMIT-REQUIRED CONFINED SPACES

☐ Evaluate all confined spaces to determine if they meet the definition of permit-required confined space.

☐ Develop a written inventory and provide a copy to all affected employees.

☐ Review the inventory at least annually to ensure accuracy.

RESPONSIBILITIES

☐ Entry Supervisor

  o Determine if a permit is needed to enter the space by referencing the confined space inventory and/or following posted warnings on spaces.

  o Know and understand the hazards within the space.

  o Know how exposure occurs, signs and symptoms of exposure, and consequences of exposure.

  o Complete the entry permit.

  o Coordinate with contract employer(s) to assure that employees of one employer do not endanger the employees of another employer.

  o Ensure that all testing as required by the permit has been conducted and results are within the range of acceptable limits and recorded appropriately on the permit.

  o Require entrants to use non-entry rescue equipment unless the equipment poses a greater threat to the safety of the entrant.

  o Ensure that all procedures and equipment specified by the permit are in place and operational.

- o Terminate entry operations when the permit has been completed or conditions not allowed under the permit arise.

- o Remove unauthorized individuals who enter or attempt to enter the space during operations.

☐ Attendant

- o Know and understand the hazards within the space.

- o Know how exposure occurs, signs and symptoms of exposure, and consequences of exposure.

- o Continuously monitor the entrants for behavioral changes due to exposure to hazards in the space.

- o Maintain an accurate count of entrants in the space at all times.

- o Remain positioned outside the confined space until relieved by another authorized attendant.

- o Inform the new attendant of all related space information (names of entrants, conditions in the space, and any other relevant information the attendant will need to perform the duties of the attendant).

- o Maintain continuous communications with all entrants.

- o Monitor activities inside and outside the space to determine if it is safe for entrants to remain in the space.

- o Know how to operate retrieval equipment should rescue be necessary.

- o Evacuate the space if a dangerous condition is detected in the space.

- o Evacuate the space if behavioral effects of hazardous exposure in the space are detected.

- o Evacuate the space if a situation occurs outside the space that could endanger the entrants.

- o Evacuate the space if he/she cannot perform all duties required of the attendant.

☐ Entrant

- o Know and understand the hazards within the space.

- o Know how exposure occurs, signs and symptoms of exposure, and consequences of exposure.

- o Properly use the equipment required for permit space entry.

- o Communicate with the attendant as necessary to allow the attendant to monitor the entrant.

- o Alert the attendant if evacuation of the space becomes necessary.

- o Warn the attendant if a prohibited condition in the space occurs.

- o Evacuate the space immediately if the attendant orders evacuation.

ENTRY PROCEDURES

☐ Secure the entry site using signs and/or barriers to prevent unauthorized entrants or visitors.

☐ Appoint a trained entry supervisor to oversee entry activities.

☐ Appoint a trained entry attendant to monitor entrants during the entire entry.

☐ Appoint only trained entrants to enter and perform work in confined spaces.

☐ Notify all affected employees in the immediate area(s) that a confined space entry is about to be conducted, the specific location of the space, and the nature of the work to be conducted.

☐ Evaluate the space for potential hazards (e.g., chemical, mechanical, physical).

☐ Disconnect, blind, or block off any hoses, pipes, or other openings which may convey flammable, injurious, or incapacitating substances into the space.

☐ Ensure inadvertent reconnection of hoses, pipes, or other openings or the removal of the blind(s) is effectively prevented.

☐ Empty, flush, or purge, through forced-air ventilation, the space of flammable, injurious, or incapacitating substances to levels that are safe for entry.

☐ If necessary to retain a safe atmosphere, continue forced air ventilation during entry operations.

☐ Lock out and dissipate any energy sources that could create a hazard in the confined space during entry.

☐ Use approved air monitoring equipment to test the atmosphere of the space for oxygen, carbon monoxide, flammability/explosives, and other hazards that may be present in the space.

☐ Ensure hot work permits are completed as required.

☐ Ensure an entry permit is completed, signed, and posted at the point of entry as required.

☐ Initiate and maintain continuous monitoring of the space to detect changes in atmospheric conditions during entry.

☐ Provide a venting means for combustion air and exhaust gases whenever oxygen-consuming equipment is used.

☐ Ensure entry and exit means are clear and protected during entry.

☐ Ensure non-entry rescue equipment is in place and ready should rescue be necessary.

☐ At conclusion of entry, account for all authorized entrants and attendants.

☐ Account for all tools, appliances, and equipment.

☐ Account for all locks, tags, blanks, blinds, or other isolating devices.

☐ Account for all personal protective equipment (PPE).

☐ Note any additional hazards identified during entry.

☐ Note any procedural changes or recommendations.

☐ Forward copies of expired permits to the appropriate department for review.

ENTRY PERMIT SYSTEM

☐ The entry permit provides a written, systematic review of hazards, communication of hazards to authorized entrants, and procedures for authorized entrants to follow when working in permit-required confined spaces.

☐ Permits will not exceed one work shift (8 or 12 hours) and will be void if the space is vacated for more than thirty (30) minutes.

☐ Atmospheric testing information must be completed for each permit-required confined space entry.

☐ List on the appropriate sections of the confined space entry permit all safety and special equipment required to enter the space.

EQUIPMENT

☐ Testing and monitoring equipment

☐ Ventilating equipment

☐ Communications equipment

☐ Personal protective equipment

☐ Lighting equipment

☐ Barriers and shields to protect entrants from outside hazards such as pedestrians or traffic

☐ Equipment, such as ladders, to enter and exit spaces safely

☐ Rescue and emergency equipment

☐ Other equipment necessary for safe entry and rescue from permit spaces

RESCUE SERVICES

☐ Rescue personnel (entry and non-entry) will be trained on the following:

- o Rescue equipment use and maintenance
- o Their respective duties during a rescue
- o Simulated rescues and entries at least once every 12 months
- o Emergency Medical Technician (EMT) Certification and/or First Aid/CPR
- o Non-entry rescue methods
- o Retrieval System Requirements (non-entry rescue):
  - ◊ A mechanical device and retrieval line must be used to retrieve personnel from vertical type permit spaces greater than five feet in depth.
  - ◊ Each authorized entrant will use a full body harness with a retrieval line attached at the center of the entrant's back, near shoulder level.
  - ◊ The retrieval line will be attached to a mechanical device or fixed point outside the permit space in such a manner that non-entry rescue can begin as soon as necessary. The mechanical device or fixed point must be capable of withstanding a dynamic stress of 5,400 pounds.

TRAINING PROGRAM

☐ General Training (All employees): During orientation and annually, all employees must be trained on the following:

- o Brief explanation of general hazards associated with confined spaces
- o Discussion of specific confined space hazards
- o Confined space identification signs and postings
- o Prohibition from entry without proper training

☐ Employees who will participate in any permit space activity will be trained in their assigned duties:

- o Prior to performing any permit space duties
- o Before there is a change in assigned duties
- o When there is a change in permit space operations that presents a hazard about which the employee has not been previously trained
- o When entry procedures have changed
- o When it is determined that the employee's knowledge or use of the program is inadequate to perform assigned duties in a safe manner.

☐ Training requirements for employees responsible for supervising, planning, entering, or participating in permit space entry and/or rescue:

- o An explanation of the general hazards associated with confined spaces
- o Duties and responsibilities as a member of a confined space entry team
- o An explanation of the permit system and other procedural requirements for conducting a confined space entry
- o Discussion of specific confined space hazards associated with the facility, location, or operation
- o Description of how to recognize probable air contaminant over-exposure symptoms to themselves and co-workers and methods for alerting attendants
- o Reason for, proper use of, and limitations of PPE and other safety equipment required for entry into permit spaces

☐ Technical Training (Entry Supervisor, Attendants, Entrants)

- o Recognition of the effects of exposure to atmospheric and chemical hazards known to be in the confined space
- o Use of air monitoring equipment and interpretation of results
- o Ventilation
- o Use of external rescue equipment
- o How to respond in emergencies

PROGRAM MONITORING

☐ This program must be reviewed on an annual basis to ensure employees understand and follow these procedures. This review should include a review of the confined space inventory, employee training records, and expired permits.

RECORDKEEPING

☐ Confined Space Inventory

☐ Written Confined Space Program

☐ Confined Space Permits

☐ Training Records

---------- END OF SECTION ----------

Section 7
Cranes (Overhead, Gantry, Crawler, & Locomotive)

OVERVIEW

This section provides information regarding the requirements to safely operate, inspect, and maintain cranes in the workplace. This section pertains to overhead, gantry, crawler, and locomotive type cranes.

REGULATORY COMPLIANCE

☐ §1910.179 Overhead and Gantry Cranes

☐ §1910.180 Crawler Locomotive and Truck Cranes

BASIC EQUIPMENT REQUIREMENTS

☐ Overhead or gantry cranes installed after August 31, 1971, must meet the specifications of the ASME (American Society of Mechanical Engineers) B30.2.

☐ The rated load capacity must be clearly identified on each side of the crane.

☐ Cranes with more than one hoisting unit will have the load capacity marked on each block.

☐ All markings must be legible from the ground or floor.

☐ All crane runways must be capable of withstanding the forces to be applied by the crane and its load.

☐ Access to the crane cabs must be by platform, stairway, or fixed ladder.

☐ Toolboxes must be securely attached to the crane and made of noncombustible materials if oils and lubricants are to be stored within.

☐ Each crane cab must contain a 10-pound fire extinguisher with an ABC rating.

☐ Cranes must be equipped with trolley stops to limit the travel of the crane trolley.

☐ Cranes must be equipped with both bridge and trolley bumpers that are energy absorbing in nature.

☐ Cranes must also be equipped with both bridge and trolley rail sweeps.

☐ All moving parts, gears, screws, keys, chains, or sprockets must be guarded if it is possible they may be contacted under normal operating conditions.

☐ All hoisting units must have at least one holding brake.

☐ Both the bridge and trolley must be equipped with a means of braking.

☐ All wiring must comply with NFPA 70—National Electric Code.

☐ Hooks must contain a latching mechanism unless the latch is made impractical by the application of the crane.

☐ All ropes, chains, and hoisting cables must meet manufacturer recommendations.

☐ Each lifting mechanism must be equipped with a hoist-limiting switch.

☐ Audible warning devices, visual warning devices, or a combination thereof must be utilized to warn that the crane is moving and/or approaching.

☐ Each crane cab must be equipped with a method of escape under emergency conditions.

BP: A centrifugal brake device with a rope that goes over the user's head and below both shoulders is recommended.

OPERATION

☐ Crane operators must be physically qualified based upon the criteria identified in ASME B30.2, Chapter 2-3.1.2.

☐ Crane operators must undergo documented training from a qualified person before being allowed to operate a crane.

BP: OSHA is expected to develop both classroom and hands-on training criteria similar to the requirements of the powered industrial truck (1910.178) standard.

☐ Crane loads must not exceed the rated capacity of the crane.

☐ A mechanical engineer must certify that occasional lifts that exceed the rated capacity of the crane can be performed in a safe manner.

☐ Crane operators must know and understand standard hand signals.

MAINTENANCE

☐ A formal preventive maintenance program as recommended by the manufacturer must be developed and followed.

☐ Lockout/Tagout procedures must be used when maintenance is performed on the crane.

☐ Signs similar to "OUT OF ORDER" must be placed on the crane during maintenance.

☐ A specific location must be provided to perform work on cranes so that hazards are not created for both maintenance and operations employees.

☐ All guards and safety devices must be replaced prior to allowing the crane back into normal operation.

INSPECTIONS (FREQUENT AND PERIODIC)

- ☐ Cranes must be frequently inspected before each use as part of a daily inspection for safe operation.

- ☐ Cranes must be periodically inspected monthly as part of a formal documented maintenance program.

- ☐ Cranes out of service for more than a year must be thoroughly inspected as required by a formal maintenance program detailed in ASME B30.2-2.1.3(b).

RECORDKEEPING

- ☐ Inspections

- ☐ Maintenance records including repair orders

- ☐ Operator medical qualification

- ☐ Operator training records

---------- END OF SECTION ----------

Section 8
Electrical Safety

OVERVIEW

The purpose of this section is to establish general safety awareness and precautions for employees who may be exposed to or working around electrical energy. OSHA's electrical standards are designed to protect employees exposed to dangers such as electric shock, electrocution, fires, and explosions. Electrical hazards are addressed in specific standards for the general industry, shipyard employment, and marine terminals. This section will cover only general industry.

REGULATORY COMPLIANCE

☐ §1910.137 Electrical Protective Devices

☐ §1910.268 Textiles (Special Industries)

☐ §1910 Subpart S, Electrical (301-399)

☐ National Fire Protection Association (NFPA) 70-1993, National Electrical Code (NEC)

☐ Local Electric Codes

☐ Equipment Manufacturers' Requirements

TRAINING/QUALIFICATIONS

☐ Training is required for employees who may be at a higher-than-normal risk of electrical shock, whether or not they are qualified to work on or near exposed, energized parts. These employees must receive appropriate classroom or on-the-job training.

☐ Only trained and qualified electrical technicians shall:

   o   Make electrical installations

   o   Repair electrical equipment

☐ Only qualified employees shall work on electric circuits, parts, or equipment, especially those that have not been de-energized and locked out. To be considered *qualified*, employees shall:

   o   Have met the training requirements above

- o Have the expertise to work safely on energized circuits
- o Understand the work practices for a specific circuit, part, or equipment, and the proper use of special precautionary techniques
- o Understand proper use of insulating and shielding materials, insulated tools, and personal protective equipment.

GENERAL REQUIREMENTS

☐ Hazardous Use Areas – Never use electrical equipment in hazardous or specially classified areas as covered in the National Electric Code unless the equipment used has the proper rating for that area.

☐ Initial Check – Before working with **any** electrical equipment, operators shall ensure it has no:
- o Frayed cords
- o Damaged cord insulation
- o Altered or damaged electrical plugs

☐ Employees Awareness – Employees must follow safe work practices when working near equipment or circuits whether or not they are energized.

☐ Arcs, Sparks Exposure – Parts of electrical equipment that produce arcs, sparks, or flames in ordinary operation must be enclosed or separated from employees and combustible material.

☐ Electrical Cords – Electrical cords must be covered in an area where they could be damaged. Examples:
- o A cord pinned beneath a cube wall or
- o On the ground subjected to vehicle or foot traffic

☐ Receptacles – Light switches, junction boxes, etc., must have the appropriate cover plate in place.

☐ Power Strips – Power strips must be plugged directly into an outlet. Do not plug into other power strips or extension cords.

☐ High-Amp Items – The following items must be plugged directly into the outlet and not a power strip or extension cord:
- o Toasters
- o Microwaves
- o Hot Plates
- o Soldering Pencils
- o Refrigerators
- o Space Heaters
- o Coil Burners
- o Hot Glue Guns
- o Heat Guns/Lamps

> **NOTE:** Flexible cords may be used with Heat Guns/Lamps and Hot Glue Guns in certain operations as long as the flexible cords used have the appropriate NEC rating for the operation.

☐ Extension Cords – Extension cords may only be used for temporary application. Never plug multiple extension cords together to make a longer cord.

☐ Conductive Clothing – Do not wear jewelry or clothing that could be conductive (watch bands, bracelets, rings, key chains, necklaces, metalized aprons, cloth with conductive thread, metal headgear) unless they are rendered nonconductive by covering, wrapping, or other insulation.

ELECTRICAL SYSTEMS – DE-ENERGIZED PARTS GUIDELINES

☐ General – All electrical circuits or parts must be de-energized before work begins on or near them, unless local management can demonstrate one of the following exceptions:

  o De-energizing could cause such additional hazards as:

    ◊ Interruption of emergency equipment

    ◊ Deactivation of alarm systems

    ◊ Elimination of, or poor lighting of area

    ◊ Shutdown of hazardous location ventilation equipment

  o De-energizing is unfeasible due to equipment or operational limitations. Examples: testing of electric circuits, work that interrupts a continuous industrial process.

  o The live circuits or parts operate at less than 50 volts to ground and there is no increased exposure from electric arcs.

☐ After De-energizing – Once the circuit or parts have been de-energized, they shall be checked to ground to ensure they have no stored energy potential, and then locked out following proper lockout procedures before continuing.

ELECTRICAL SYSTEMS – ENERGIZED PARTS GUIDELINES

☐ Procedures – If the exposed live parts cannot or should not be de-energized, employees still must be protected from contact with energized circuits or parts directly or indirectly through a conductive object by specific procedures. These shall address:

  o The conditions under which the work is to be performed

  o The voltage level of the exposed electric conductors or circuit parts

☐ Employees shall be familiar with:

  o The particular work practices regarding that specific circuit, part, or equipment, and the proper use of special precautionary techniques.

  o The proper use of shielding materials, insulated tools, and personal protective equipment.

PROTECTIVE EQUIPMENT GUIDELINES

☐ Responsibility - Employees working in areas with potential electrical hazards shall use the electrical protective equipment appropriate to the specific parts of the body to be protected and for the specific work to be performed.

☐ Personal Protective Equipment (PPE)

  o Employees shall wear nonconductive head protection to protect against head injury caused by electric shock or burns from exposed energized parts.

  o Employees shall wear eye and face protection to protect the eyes or face from electric arcs or flashes or from flying objects due to electrical explosion.

o Protective shields, protective barriers, or insulating materials shall protect employees working near exposed energized parts from shock, burns, or other electrically related injuries that might result from accidental contact, dangerous electric heating, or arcing. When normally enclosed live parts are exposed for maintenance or repair, they shall be guarded to protect persons from contact.

☐ Insulation Protection

o If the protective equipment's insulation may be damaged during use, the insulating material shall be protected. (For example, an outer covering of leather is sometimes used to protect rubber-insulating material.)

o When working near exposed energized conductors or circuit parts, employees shall use insulated tools or handling equipment if the tools or handling equipment might make contact with such conductors or parts. If the insulation of the tools or handling equipment is subject to damage, the insulating material shall be protected.

o Fuse handling equipment, insulated for the current voltage, shall be used to remove or install fuses when the fuse terminals are energized.

o Conductive materials and equipment in contact with any part of an employee's body shall be handled to prevent them from contacting exposed energized conductors or circuit parts.

☐ General Protective Guidelines

o Portable ladders shall have nonconductive side rails if they are used where employees or the ladder could contact exposed parts.

o Ropes and hand lines near exposed energized parts shall be nonconductive and dry.

o A nonconductive covering or wrapping or insulation shall cover conductive articles of jewelry or clothing (such as watch bands, bracelets, rings, key chains, necklaces, metalized aprons, cloth with conductive thread, or metal headgear) or such articles shall not be worn when working with energized parts.

ALERTING TECHNIQUES

☐ Use safety signs, safety symbols, or accident prevention tags to warn employees about electrical hazards.

☐ Use barricades with safety signs to limit access to work areas with non-insulated energized conductors or circuit parts.

☐ Do not use conductive barricades where they might cause an electrical hazard.

☐ If signs and barricades do not provide sufficient warning and protection from electrical hazards, an attendant shall be stationed to warn and protect employees.

LIGHTING AREAS

☐ Employees shall not enter spaces with exposed energized parts without proper lighting to enable employees to identify all potential hazards.

☐ Where there is poor lighting or obstruction, employees shall not perform tasks near exposed energized parts.

☐ Employees shall not reach blindly into areas that may contain energized parts.

HOUSEKEEPING REQUIREMENTS FOR WORK ON ENERGIZED PARTS

☐ Hazardous Areas

- o Employees shall not enter spaces with exposed energized parts without lighting that enables those employees to identify all potential hazards.

- o Employees shall not perform tasks near exposed energized parts where there is poor lighting or other eyesight obstructions.

- o Employees shall not reach blindly into areas that may contain energized parts.

☐ Electrically conductive cleaning materials (steel wool, metalized cloth, silicon carbide, and conductive liquids) shall not be used near energized parts unless written procedures are developed and followed to prevent electrical contact.

☐ The minimum clearances for work on energized equipment shall be:

- o 36 inches for 0 to 150 volts

- o 48 inches for 151 to 600 volts

> **NOTE:** Normal clearance in front of transformers and circuit breaker panels shall be 36 inches; this area shall not be used for storage at any time.

STATIC ELECTRICITY

☐ Caused by friction, static electricity can be a hazard when a spark discharges near flammable mixtures or if an employee completes the circuit and the static electricity discharges through him. Be aware that the presence of static electricity is not always apparent until a discharge or spark occurs.

☐ Causes

- o Wool and nylon sliding over each other in low-humidity conditions

- o Fluid flowing through a hose

- o Moving helicopter blades

- o Walking on certain rugs

- o Air flowing over aircraft surfaces

- o Actions involving friction such as buffing, wiping, or stripping

- o Induction from an electrically charged atmosphere

☐ Ground objects to safely discharge electrical buildup.

☐ Bond objects to each other with a conductor to prevent accumulation of static electricity between the objects.

☐ Wear anti-static or cotton clothing. Do not wear rayon, nylon, silk, wool, or plastic clothing.

☐ Marking Equipment

- o Electrical equipment should have clear, durable markings of manufacturer's name and trademark, voltage, current, wattage, or any other ratings.

- o Mark each disconnecting means, over-current device, switch, or other control device to indicate its purpose, unless located or arranged so its purpose is evident.

o   Equipment requiring clearances under the "Housekeeping Requirements for Work on Energized Parts" subsection shall have those clearance areas marked in yellow.

o   All distribution equipment (controls, junction boxes, emergency disconnects, etc.) should also have the above clearance areas marked in yellow so they remain accessible at all times

☐ Accessibility – Each disconnecting means, over-current device, switch, or other control device shall be accessible at all times for an emergency.

> **NOTE:** "Accessible" means workers can reach it without moving items, climbing over items, or otherwise being delayed. Mark distribution equipment so it is not blocked by storage items, etc.

EMERGENCY ACTIONS FOR ELECTRIC SHOCK

☐ Follow these procedures when a person comes in contact with a live electrical conductor or is hit by lightning:

o   Remove Contact – If the victim is still in contact with the current, remove him from the current *immediately* following the procedures below.

o   Power – Turn off power if possible

o   Pushing – Knock or push the victim away from the conductor using non-conducting material such as dry wood, PVC pipe, etc.

o   Pulling – Drag the victim away from the conductor using dry clothing, leather belt, etc.

o   Call for Assistance – After the victim is free from the current, coordinate medical assistance for him.

> **NOTE:** This may include performing CPR or other medical treatment at the level for which one is qualified.

----------END OF SECTION----------

Section 9
Emergency Action (Contingency) Planning

OVERVIEW

The purpose of this section is to assist with the development and maintenance of the written emergency action (or contingency) plan. It is the responsibility of the employer to ensure that all employees are thoroughly trained and ready for all emergencies. The types of emergencies selected should consider the demographics of the facility (e.g., include planning for tornados and hurricanes in Southeast Texas, earthquakes for California, community hazard communication for potentially dangerous operations within heavily populated areas, etc.). Also include other types of emergency response for bomb or terror threats.

REGULATORY COMPLIANCE

☐ §1910.38 Emergency Action Plans

GENERAL REQUIREMENTS

☐ Mark all routes of escape throughout the facility (specific markings are discussed in the Exits section).

BP: In addition to emergency lighting and exit signs, it is highly recommended to mark lanes throughout the facility to readily identify the appropriate lanes of travel.

☐ Determine each type of emergency that may occur at the facility. For example, the most common or recurring emergency may be a fire or hazardous chemical spill. But, be well aware of geological or "Act of God" emergency possibilities in your specific climate such as the threat of tornadoes, earthquakes, hurricanes, and floods. Also, consider the type of location the facility is in (urban or rural) and your specific type of business. It may be necessary to prepare for malicious acts such as riots or bomb threats.

☐ Develop a written plan to include preparation and training for the above and an action sequence in the event of an emergency.

☐ Test emergency response equipment on a monthly basis, including all alarms, emergency lighting, battery-backed exit signs, and fire extinguishers.

☐ Perform annual fire drills for evacuation procedural training.

WRITTEN EMERGENCY ACTION PLAN

- ☐ Include emergency escape procedures and route assignments.

- ☐ Procedures for employees who remain to operate critical plant operations before they evacuation.

- ☐ Procedures for accounting for all employees after evacuation.

- ☐ Rescue and medical instructions for employees who are to perform them (first-aid trained personnel, company nurse, etc.).

- ☐ The preferred means of reporting emergencies.

- ☐ Names or regular job titles of personnel that can be contacted for further information or explanation of duties under the plan.

- ☐ Procedures for handling each determined emergency possibility.

EMERGENCY ESCAPE PROCEDURES

- ☐ Designate a central location for employees to gather for a head count.

- ☐ Develop and place diagrams of the facility in conspicuous locations throughout that show the nearest exit to that location.

- ☐ Post a list of emergency telephone numbers by every telephone or communication system (radio).

- ☐ Predetermine routes or means of egress throughout the facility and mark accordingly. Keep these routes free from debris and blockage.

BP: The central location for employees to gather is called a "meet point" or "rally point." Mark these rally points where they can be seen upon exiting the building. Pre-assign a person or persons (as a site coordinator) in charge of these areas to facilitate each. Their purpose is to account for personnel. If there are multiple departments or areas that will be gathering at a particular rally point, assign responsibility to the supervisors to assist the site coordinator.

It may be necessary to have several rally points for a large facility. Plan accordingly with radios, cell phones, etc. to be able to communicate headcount back to a central point. Not being able to do so could put fire department personnel at unnecessary risk since they must assume that any missing or unaccounted personnel are still inside.

Site coordinators should maintain order at rally points by not allowing employees to leave unless and until the emergency response manager or fire department authority has given permission to do so. The site coordinator should also keep employees out of the way of fire or police activities and remain behind fire lanes. No employees should be allowed to re-enter the building until the emergency has been cleared and permission to re-enter granted by a fire department official.

ALARM SYSTEM

- ☐ Different distinct signal for each emergency type (hearing disabled employees shall be provided with visual alarm systems).

- ☐ Test every month for correct working operation and audibility.

TRAINING

☐ All employees shall be trained in the proper emergency and evacuation plans for the area where they will be working. Some areas to focus on include:

- o Location of emergency numbers and how/where to report emergencies
- o Proper procedures for evacuation and the routes to take
- o Recognition and reaction to chemical and fire hazards
- o Recognition of the various emergency alarms
- o Communication during emergencies
- o Responsibilities asked of the employee
- o Location of fire fighting equipment and how to use it
- o Emergency first aid and CPR for key employees
- o Training should be provided at least annually

RECORDKEEPING

☐ Training Documentation

☐ Written Emergency Evacuation Plan

☐ Inspection Records (for alarms, emergency lighting, and fire extinguishers).

---------- END OF SECTION ----------

Section 10
Employee Records (Exposure and Medical)

OVERVIEW

This section provides information regarding the access for employees and/or their representatives to relevant exposure and medical records.

REGULATORY COMPLIANCE

☐ §1910.1020 Access to Employee Exposure and Medical Records

ACCESS REQUIREMENTS

☐ Any employee, current or former, and his/her representative has the right to review and/or receive a copy of any record of employee exposure or medical records, including medical analysis.

☐ Maintain a copy of the standard governing access (§1910.1020 Access to Employee Exposure and Medical Records).

☐ A request from an employee representative shall be in writing and shall specify the particular records requested and the occupational health need for gaining access (See Figure 1 on the following page for a sample).

☐ The employer may limit access to such information as should be readily known to the requester and which may be necessary to locate and identify the records requested (e.g., dates and locations where employee worked during time in question).

☐ Copies of records should be provided without costs to the employee.

☐ Employers may charge a reasonable fee for subsequent requests of the previously provided copies.

Figure 1: Sample Letter Authorizing Release of Employee Medical Record Information to a Designated Representative

I, _____ (full name of worker/patient), hereby authorize _____ (individual or organization holding the medical records) to release to _____ (individual or organization authorized to receive the medical information), the following medical information from my personal medical records:

_____

_____

(Describe generally the information desired to be released)

I give my permission for this medical information to be used for the following purpose:

_____

_____

but I do not give permission for any other use or re-disclosure of this information.

Full name of Employee or Legal Representative: _____

Signature of Employee or Legal Representative: _____

Date of Signature: _____

NOTE: You may want to leave lines for any further information such as adding an expiration date for this request or for instructions on portions of these records you do not intend to be released.

MEDICAL RECORDS

☐ Maintain all medical records for the duration of employment plus 30 years.

☐ Audiometric testing (baseline and annual audiograms).

☐ Chest x-ray (available for review, but do not have to be loaned or copied).

☐ Descriptions of treatments.

☐ Employee medical complaints.

☐ Post-employment physical.

☐ Pre-employment physical.

EXPOSURE RECORDS

☐ Maintain exposure records for the duration of employment plus 30 years.

☐ Air measurement or monitoring results (Industrial Hygiene sampling).

☐ Similar samples of other past or present employees if the above is not available.

☐ Records that indicate the amount and nature of toxic substances at the workplace or to which the employee is being assigned or transferred.

☐ Noise monitoring results.

BP: Ergonomic measurements that may indicate a repetitive stress or other musculoskeletal disorders should also be maintained.

OTHER RECORDS (Must be made available to the employee)

☐ Measures for controlling worker exposure to chemicals (job safety analysis that identifies required personal protective equipment, ventilation, material handling procedures, etc.).

☐ Methodologies used to gather data (types of testing or monitoring devices used, procedures, areas included, and substances monitored such as vapors, fumes, gases, or dusts).

EMPLOYER RIGHTS AND RESPONSIBILITIES

☐ The employer shall provide access to records within 15 days of the request or provide the requester a reason for delay and the earliest date the record can be made available.

☐ The employer shall not charge for the first copy or any additional information at another time.

☐ The employer may charge a reasonable, nondiscriminatory price (e.g., search and copying expenses, but not overhead expenses) for a second copy of the same information.

☐ If a copy machine is not available, the documents may be loaned for a reasonable time to have copies made.

> **NOTE:** This is not recommended. Send office personnel or other associate to have copies made without releasing originals.

☐ Delete names and identifiers of employees, such as height, sex, weight, etc., when records, x-rays, etc. are sent for analysis.

☐ X-rays may be loaned at employer discretion, but viewing in-house is sufficient and recommended.

TRAINING

☐ Initial employment and annual recurrent—the existence, location, types of records, and person to contact to retrieve information

☐ The procedure for accessing records

☐ The right to access medical records

---------- END OF SECTION ----------

Section 11
Exit Routes

OVERVIEW

This section provides specific information regarding the proper maintenance, safeguarding, and operational features for exit routes. *Exit route* means a continuous and unobstructed path of exit travel from any point within a workplace to a place of safety (including refuge areas). An exit route consists of three parts: The exit access; the exit; and the exit discharge. (An exit route includes all vertical and horizontal areas along the route.) The designation of refuge or safe areas for evacuation should be determined and identified in the emergency action plan. In a building divided into fire zones by fire walls, the refuge area may still be within the same building but in a different zone from where the emergency occurs.

REGULATORY COMPLIANCE

☐ §1910.34 Coverage and Definitions

☐ §1910.35 Compliance with NFPA 101-2000, Life Safety Code

☐ §1910.36 Design and Construction Requirements for Exit Routes

☐ §1910.37 Maintenance, Safeguards, and Operational Features for Exit Routes

☐ National Fire Protection Association (NFPA) 101: Life Safety Code 2006 Edition

GENERAL REQUIREMENTS FOR DESIGN AND CONSTRUCTION OF EXIT ROUTES

☐ An exit must be separated by fire resistant materials. Construction materials used to separate an exit from other parts of the workplace must have a one-hour fire-resistance rating if the exit connects three or fewer stories and a two-hour fire-resistance rating if the exit connects four or more stories.

☐ Openings into an exit must be limited. An exit is permitted to have only those openings necessary to allow access to the exit from occupied areas of the workplace, or to the exit discharge. An opening into an exit must be protected by a self-closing fire door that remains closed or automatically closes in an emergency upon the sounding of a fire alarm or employee alarm system.

☐ Two exit routes – At least two exit routes must be available in a workplace to permit prompt evacuation of employees and other building occupants during an emergency. The exit routes must be located as far away as practical from each other so that if one exit route is blocked by fire or smoke, employees can evacuate using the second exit route. *Exception:* A single exit route is permitted where the number of employees, the size of the building, its occupancy, or the arrangement of the workplace is such that all employees would be able to evacuate safely during an emergency.

- [ ] <u>More than two exit routes</u> – More than two exit routes must be available in a workplace if the number of employees, the size of the building, its occupancy, or the arrangement of the workplace is such that all employees would not be able to evacuate safely during an emergency.

- [ ] <u>Exit discharge</u> – Each exit discharge must lead directly outside or to a street, walkway, refuge area, public way, or open space with access to the outside.

- [ ] The street, walkway, refuge area, public way, or open space to which an exit discharge leads must be large enough to accommodate the building occupants likely to use the exit route.

- [ ] Exit stairs that continue beyond the level on which the exit discharge is located must be interrupted at that level by doors, partitions, or other effective means that clearly indicate the direction of travel leading to the exit discharge.

- [ ] An exit door must be unlocked. Employees must be able to open an exit route door from the inside at all times without keys, tools, or special knowledge. A device such as a panic bar that locks only from the outside is permitted on exit discharge doors.

- [ ] Exit route doors must be free of any device or alarm that could restrict emergency use of the exit route if the device or alarm fails.

- [ ] An exit route door may be locked from the inside only in mental, penal, or correctional facilities and then only if supervisory personnel are continuously on duty and the employer has a plan to remove occupants from the facility during an emergency.

- [ ] A door that connects any room to an exit route must swing out in the direction of exit travel if the room is designed to be occupied by more than 50 people or if the room is a high hazard area (e.g., contains contents that are likely to burn with extreme rapidity or explode).

CAPACITY AND SIZE OF EXIT ROUTES

- [ ] Exit routes must support the maximum permitted occupant load for each floor served. The capacity of an exit route may not decrease in the direction of exit route travel to the exit discharge.

- [ ] The ceiling of an exit route must be at least 7 feet 6 inches (2.3 m) high. Any projection from the ceiling must not reach a point less than 6 feet 8 inches (2.0 m) from the floor.

- [ ] An exit access must be at least 28 inches (71.1 cm) wide at all points. Where there is only one exit access leading to an exit or exit discharge, the width of the exit and exit discharge must be at least equal to the width of the exit access.

- [ ] The width of an exit route must be sufficient to accommodate the maximum permitted occupant load of each floor served by the exit route.

- [ ] Objects that project into the exit route must not reduce the width of the exit route to less than the minimum width requirements for exit routes.

OUTDOOR EXIT ROUTE

☐ An outdoor exit route is permitted. The outdoor exit route must have guardrails to protect unenclosed sides if a fall hazard exists. The outdoor exit route must be covered if snow or ice is likely to accumulate along the route, unless the employer can demonstrate that any snow or ice accumulation will be removed before it presents a slipping hazard. The outdoor exit route must be reasonably straight and have smooth, solid, substantially level walkways. The outdoor exit route must not have a dead-end that is longer than 20 feet (6.2 m).

EXIT ROUTE HAZARDS

☐ The danger to employees must be minimized. Exit routes must be kept free of explosive or highly flammable furnishings or other decorations.

☐ Exit routes must be arranged so that employees will not have to travel toward a high hazard area, unless the path of travel is effectively shielded from the high hazard area by suitable partitions or other physical barriers.

☐ Exit routes must be free and unobstructed. No materials or equipment may be placed, either permanently or temporarily, within the exit route. The exit access must not go through a room that can be locked, such as a bathroom, to reach an exit or exit discharge, nor may it lead into a dead-end corridor. Stairs or a ramp must be provided where the exit route is not substantially level.

☐ Safeguards designed to protect employees during an emergency (e.g., sprinkler systems, alarm systems, fire doors, exit lighting) must be in proper working order at all times.

LIGHTING AND MARKING

☐ Lighting and marking must be adequate and appropriate. Each exit route must be adequately lighted so that an employee with normal vision can see along the exit route.

☐ Each exit must be clearly visible and marked by a sign reading "EXIT".

☐ Each exit route door must be free of decorations or signs that obscure the visibility of the exit route door.

☐ If the direction of travel to the exit or exit discharge is not immediately apparent, signs must be posted along the exit access indicating the direction of travel to the nearest exit and exit discharge. Additionally, the line-of-sight to an exit sign must clearly be visible at all times.

☐ Each doorway or passage along an exit access that could be mistaken for an exit must be marked "NOT AN EXIT" or similar designation, or be identified by a sign indicating its actual use (e.g., closet).

☐ Each exit sign must be illuminated to a surface value of at least 5 foot-candles (54 lux) by a reliable light source and be distinctive in color. Self-luminous or electroluminescent signs that have a minimum luminance surface value of at least .06 foot-lamberts (0.21 cd/m^2) are permitted.

☐ Each exit sign must have the word "EXIT" in plainly legible letters not less than 6 inches (15.2 cm) high, with the principal strokes of the letters in the word "EXIT" not less than 3/4 inch (1.9 cm) wide.

☐ The fire retardant properties of paints or solutions must be maintained. Fire retardant paints or solutions must be renewed as often as necessary to maintain their fire retardant properties.

EXIT PROTECTION DURING CONSTRUCTION

☐ During new construction, employees must not occupy a workplace until the exit routes required by this section are completed and ready for employee use for the portion of the workplace they occupy.

☐ During repairs or alterations, employees must not occupy a workplace unless the exit routes required by this section are available and existing fire protections are maintained, or until alternate fire protection is furnished that provides an equivalent level of safety.

☐ Employees must not be exposed to hazards of flammable or explosive substances or equipment used during construction, repairs, or alterations that are beyond the normal permissible conditions in the workplace, or that would impede exiting the workplace.

EMPLOYEE ALARM SYSTEM

☐ An employee alarm system must be operable. Employers must install and maintain an operable employee alarm system that has a distinctive signal to warn employees of fire or other emergencies, unless employees can promptly see or smell a fire or other hazard in time to provide adequate warning to them. The employee alarm system must comply with §1910.165.

TERMS/DEFINITIONS

☐ **_Exit_** – The portion of an exit route that is generally separated from other areas to provide a protected way of travel to the exit discharge. An example of an exit is a 2-hour fire-resistance rated enclosed stairway that leads from the fifth floor of an office building to the outside of the building.

☐ **_Exit access_** – The portion of an exit route that leads to an exit. An example of an exit access is a corridor on the fifth floor of an office building that leads to a 2-hour fire-resistance rated enclosed stairway (the Exit).

☐ **_Exit discharge_** – The part of the exit route that leads directly outside or to a street, walkway, refuge area, public way, or open space with access to the outside. An example of an exit discharge is a door at the bottom of a 2-hour fire-resistance rated enclosed stairway that discharges to a place of safety outside the building.

☐ **_Exit route_** – A continuous and unobstructed path of exit travel from any point within a workplace to a place of safety (including refuge areas). An exit route consists of three parts: the exit access; the exit; and, the exit discharge. (An exit route includes all vertical and horizontal areas along the route.)

☐ **_High hazard area_** – An area inside a workplace in which operations include high hazard materials, processes, or contents.

☐ **_Occupant Load_** – The total number of persons that may occupy a workplace or portion of a workplace at any one time. The occupant load of a workplace is calculated by dividing the gross floor area of the workplace or portion of a workplace by the occupant load factor for that particular type of workplace occupancy. Information regarding "Occupant load" is located in NFPA 101-2000, Life Safety Code.

☐ **_Refuge Area_** – Either:

    o   A space along an exit route protected from the effects of fire by separation from other spaces within the building by a barrier with at least a 1-hour fire-resistance rating; or

    o   A floor with at least two spaces, separated from each other by smoke-resistant partitions, in a building protected throughout by an automatic sprinkler system that complies with §1910.159.

☐ **_Self-luminous_** – A light source that is illuminated by a self-contained power source (e.g., tritium) and that operates independently from external power sources. Batteries are not acceptable self-contained power sources. The light source is typically contained inside the device.

---------- END OF SECTION ----------

Section 12
Fire Extinguishers

OVERVIEW

The requirements of this section apply to the placement, use, maintenance, and testing of portable fire extinguishers provided for the use of employees. Selection and distribution requirements of this section do not apply to extinguishers provided for employee use on the outside of workplace buildings or structures. Where extinguishers are provided but are not intended for employee use, and the employer has an emergency action plan and a fire prevention plan that meet the requirements of 29 CFR 1910.38 and 29 CFR 1910.39 respectively, then only the requirements for maintenance, inspection, and testing apply.

REGULATORY COMPLIANCE

☐ §1910.157 Portable Fire Extinguishers

☐ National Fire Protection Association (NFPA) Code 10

GENERAL REQUIREMENTS

☐ Small fires may easily be extinguished with the proper fire extinguishers. Extinguishers can represent an important segment of any fire protection program; however, successful use depends on the following requirements:

  o The extinguisher is properly located, in working order, and of the proper type for a fire that may occur.

  o The fire is discovered and addressed while still small enough for the extinguisher to be effective.

  o The fire is addressed by a person familiar with the proper use of the extinguisher.

BP: The placement of fire extinguishers should be for use on incipient (early) stage fires or to assist in emergency egress only. They are not for battling established fires. If the fire can be extinguished safely with fire extinguishers in a few moments, without endangering employees or others, the employee may use fire extinguishers (if applicable to employer's policy and the employee has been trained to do so). Otherwise, the employer's policy should mandate the employee's responsibility to sound the alarm and evacuate the area.

☐ Extinguisher Maintenance: Extinguishers shall be maintained in a fully charged and operable condition at all times in all areas where it is possible for someone to use extinguishers. This shall include open shelves in storerooms. If the extinguisher is not operable, then it shall be replaced, tagged "not operable," removed from service, and stored so that no one will attempt to use it in an emergency.

☐ Extinguisher Locations: Extinguishers shall be conspicuously located where they will be readily accessible and immediately available in case of fire. They shall be located along normal paths of travel. The number and size of extinguishers shall conform to the specifications outlined in this Section.

Note: Extinguishers must never be obstructed or blocked.

☐ Inspection Tags: All extinguishers shall have current inspection tags that conform to local regulations at all times while placed in an area where someone could attempt to use extinguishers. This shall include open shelves in storerooms. Inspection shall conform to the requirements outlined in this Section.

☐ Marking Locations: Extinguishers shall not be obstructed or obscured from view. In locations where visual obstruction is not avoidable, means shall be provided to indicate the location of the extinguisher (e.g., painting a red band around a column or affixing a label reading "Extinguisher" and pointing to its location). This does not apply to wheeled fire extinguishers; however, wheeled extinguishers must remain unobstructed.

☐ Different Classes: If extinguishers intended for different classes of fire are grouped together, their intended use shall be marked conspicuously to ensure the proper choice of extinguisher for fire suppression.

☐ Mounting: Extinguishers shall be mounted in positions that are easily accessible and shall not subject employees to injury.

☐ Operating Instructions: Extinguishers mounted in cabinets, wall recesses, or set on shelves shall be placed so that operating instructions face outward.

☐ Vibrating Locations: Extinguishers installed under conditions subject to severe vibration shall be installed in brackets specifically designed to handle vibration.

☐ Temperature Range: Extinguishers shall be able to operate in weather conditions up to 120°F. Water-type extinguishers shall not be installed in locations where temperatures may be below 40°F. When extinguishers are installed in locations subjected to temperatures outside the range prescribed above, they shall be of a type approved for the temperature exposure.

CLASSIFICATIONS OF FIRES

☐ **Class A Fires:** Fires involving combustible materials such as wood, cloth, paper, rubber, or many plastics.

Class A
• Ordinary Combustibles

☐ **Class B Fires:** Fires involving flammable liquid, gases or greases, oils, tars, oil-based paints, or lacquers.

Class B
• Flammable Liquids and Gases

☐ **Class C Fires**: All types of fires that involve live electrical equipment where the electrical non-conductivity of the extinguishing media is of importance. (After electrical equipment is de-energized, the fire becomes a Class A or Class B fire.)

☐ **Class D Fires**: Fires involving combustible metals, such as magnesium, titanium, zirconium, sodium, lithium, or potassium.

DISTRIBUTION OF FIRE EXTINGUISHERS

☐ Number/Location: The number of fire extinguishers needed to protect a property shall be determined by considering the area and arrangement of the building or occupancy, the severity of hazard, the anticipated classes of fires, and the distances to be traveled to reach the extinguishers.

☐ Office Areas: Office areas normally require protection against Class A hazards. A fire extinguisher with a minimum rating of 2-A is needed for every 6,000 square feet of floor area and spaced so that the maximum travel distance from any point in the space protected by the extinguisher to the extinguisher location will not exceed 75 feet. Each office area (an "area" is defined as the space located within a set or sets of fire doors) shall be provided with a minimum of one extinguisher, regardless of square footage.

☐ Hazard Areas:

- o Ordinary Hazard Areas: Protection for ordinary hazard areas is provided by installing as a minimum one 2A:20B:C extinguisher for each 3,000 square feet of floor area, provided travel distance within that area to the extinguisher location does not exceed 50 feet. If preferred, in lieu of the above, one 4A:40B:C extinguisher can be used for each 6,000 square feet of floor area with a maximum travel distance of 50 feet. Each shop, regardless of area, shall have a minimum of one 2A:20B:C extinguisher.

- o High Hazard Areas: Examples include woodworking shops, warehouses with combustibles piled over 12 feet, and processes such as flammable liquid handling, painting, etc. Protection against these hazards is provided by installing as a minimum one 4A:40B:C extinguisher for each 4,000 square feet of floor area provided travel distance does not exceed 30 feet or one 4A:80B:C with a maximum travel distance of 50 feet.

- o Flammable Liquid Areas: For areas that have dip tanks or other tanks of flammable liquids, Class B fire extinguishers shall be provided.

☐ Metal Fires: Metal fires are classified as Class D fires (combustible metal fires). Class D extinguishers or containers of extinguishing media shall be kept within easy reach of each operator performing a machine operation on magnesium or other metal that is subject to fire.

INSTALLATION OF PERMANENTLY LOCATED EXTINGUISHERS

☐ Location, Access, and Installation

- o Location: Extinguishers shall be conspicuously located along normal paths of travel, including exits.

- o Access: Extinguishers shall be assigned to permanent locations so that easy access is maintained at all times.

- o Markings: Extinguishers shall not be obscured from view or obstructed. Extinguishers should have clearance of 18 inches on each side. In areas where the extinguisher cannot be readily visible, markings are installed as follows:

  - ◊ Subject to local regulations, extinguishers mounted on interior columns must have a red-colored band at least 12 inches wide painted around the 4 sides of the column. The band shall be located high enough on the column to be readily visible from all directions.

  - ◊ Extinguishers shall be installed on the hangers or brackets supplied, mounted in cabinets, or set on shelves, unless the extinguishers are of the wheeled type.

  - ◊ If mounted in cabinets or wall recesses or set on shelves, the extinguishers shall be placed in a manner so that the operating instructions face outward. Signage for fire extinguisher location shall project from the wall to enhance lateral visibility. If location is subject to severe vibration, the extinguisher shall be installed in brackets specifically designed to cope with this vibration.

☐ Weight Factors: Extinguishers weighing 40 pounds or less when filled shall be installed so that the top of the extinguisher is no more than 5 feet above the floor. Extinguishers weighing more than 40 pounds (except wheeled units) shall be installed so that the top of the extinguisher is no more than 3½ feet above the floor. In no case shall the clearance between the bottom of the extinguisher and the floor be less than 4 inches.

INSPECTION, MAINTENANCE, AND SERVICE OR REPLACEMENT

☐ All fire extinguishers must be inspected on a monthly basis. This inspection may be done internally or externally.

BP: Although how and where to document inspections is not specified in the standard, it is recommended to keep the record as an attached tag with the date of the monthly inspection and initials of the person performing the inspection clearly included on the tag.

☐ Any time fire-fighting equipment is used or missing from its assigned location, notify the proper personnel to ensure rapid replacement.

☐ If an extinguisher is removed for service, it shall be replaced immediately with a spare extinguisher, if available.

INSPECTION GUIDELINES

☐ Annual inspections shall be performed by an outside contractor.

☐ For monthly inspections:

- o Check that the extinguisher is in a designated place and operating instructions on the nameplate are legible and facing outward.

- o Examine the nozzle and hose for obvious physical damage, cracking, or clogs.

o Check seal, wire, and gauge, if applicable. Both the seal and wire must be intact and installed in such a manner that the extinguisher cannot be used without breaking the seal.

o Check the accessibility and security of each extinguisher. In the case of wall-mounted extinguishers, make certain that aisles are open or, in the case of cart type, that they are strategically located. Ensure that each extinguisher is secure in its brackets.

o Check the record card to ensure that it is the correct card, current, and securely attached to the extinguisher.

o Check the general condition of the complete extinguisher. See that no damage has occurred that will render the extinguisher inoperative.

o For wheeled units, rotate the wheels to ensure they rotate freely; lubricate if necessary.

o Record the inspection as outlined in this Section.

HYDROSTATIC TESTS

☐ Hydrostatic tests of fire extinguishers are based on the type of fire extinguisher and shall be performed by an outside contractor.

☐ Hydrostatic tests are required for all fire extinguishers at given intervals and performed by an outside contractor. See Table 1 below for a schedule of required intervals. If at any time an extinguisher shows evidence of corrosion or mechanical damage, it shall be subjected to a hydrostatic test and/or replaced.

Extinguisher Type	Test Interval (years)
Stored-Pressure Water, Loaded Stream, and/or Antifreeze	5
Dry Chemical with stainless steel shell	5
Carbon Dioxide	5
Dry Chemical, stored pressure	12
Halon 1211/1301	12

Table 1

TRAINING AND QUALIFICATIONS

☐ All personnel who are expected to use fire-fighting equipment shall be trained to properly use that equipment and be familiar with its location.

☐ Only employees who have been trained and are familiar with the general principles of fire extinguisher use and with the hazards involved with incipient (early) stage fire fighting shall attempt to use fire extinguishers.

BP: This training need not include actual hands-on fire fighting training unless required by local fire code.

☐ Training Format: Familiarization training shall be provided upon initial assignment to an area that may require use of a fire extinguisher and at least annually thereafter if an employee remains in the assignment.

☐ Inspection: Employees assigned the task of fire system inspection shall be familiar with the requirements of this section and applicable local codes for fire extinguishers.

☐ Maintenance: Employees assigned to fire extinguisher maintenance shall be trained to recognize problems with fire extinguishers and trained on proper maintenance procedures.

RECORDKEEPING

☐ Training Documentation: All training in fire-fighting and use of portable fire extinguishers shall be documented and kept on file until recurrent training takes place.

☐ Monthly Extinguisher Inspection: Record the date of inspection and the initials of the person performing the inspection.

☐ Annual Extinguisher Inspection: Record the month and year the inspection was performed and the identity of the person and agency performing the service. The information shall be recorded on a tag or similar label securely attached to the fire extinguisher until replaced by the next annual inspection tag (provided by contractor).

☐ Hydrostatic Inspection/Test: A certified record of the testing that includes the serial number or other identifier of each extinguisher in the facility, date of hydrostatic testing, and the signature of the person performing the test shall be maintained and readily available for purposes of inspection.

---------- END OF SECTION ----------

Section 13
Fire Prevention

OVERVIEW

This section describes the regulations and best practices for the prevention of fires and also outlines emergency action preparation measures. Although some of the information in this section is regulatory-oriented, most of the information is a best practice (**BP**) for facilities that do not utilize fire brigades. OSHA regulations for general industry do not specifically include fire prevention techniques, but do cover areas for fire protection. In this text, there is a separate section that addresses Fire Protection.

REGULATORY COMPLIANCE

☐ §1910.252 General Requirements (Welding, Cutting, and Brazing)

☐ §1910 Subpart L, Fire Protection

☐ National Fire Protection Association (NFPA) 70, National Electric Code (NEC)

TRAINING AND QUALIFICATIONS

☐ All employees who are expected to use firefighting equipment shall be trained to properly use that equipment and be familiar with its location.

☐ Only employees who have been trained and are familiar with the general principles of fire extinguisher use and with the hazards involved with incipient (early) stage fire fighting shall attempt to use fire extinguishers; all others are to sound the alarm and evacuate the area.

> **NOTE:** An employer can have a policy of "alarm and flee" where the employer is not required to train employees in the use of fire extinguishers, but must train them to sound the alarm and evacuate the area. This is not a recommended policy to take since the training for fire extinguisher use takes little more time than training to "alarm and flee." It could also be very difficult to ensure compliance since some employees may have the tendency to use the extinguisher anyway. In addition, if an OSHA officer inspects the facility and asks an employee "What would your reaction be to a small fire?", an answer that includes using a fire extinguisher would indicate non-compliance with company policy and the regulations.

EMERGENCY ACTIONS

Anyone noticing a fire, large or small, should observe the following rules:

☐ Sound the alarm. Alert **all** persons in the vicinity of the fire.

☐ Evacuate the area. If the fire is in an area of possible explosion, seek shelter or get clear of the danger zone.

☐ Call the fire department. Call the posted numbers to be used in the event of a fire.

☐ Attempt to control the fire. If the fire is in its early or incipient stage, and you have been trained to do so, use available equipment and material to control, minimize, or eliminate the fire. If the fire has been extinguished prior to the fire department's arrival, request that the fire department check for hidden embers or other concealed danger signs. Remember, only trained personnel shall use fire extinguishing equipment.

> **NOTE:** See the Fire Extinguisher Section for specific fire extinguisher information and use.

GENERAL RULES

Observe the following general rules applying to fire prevention procedures:

☐ Sources of Ignition – Sources of ignition must be strictly controlled when spraying flammable materials or when flammable vapors are present.

   o Do not operate gasoline-powered equipment, electrical switches, drills, and motors (unless explosion-proof), or anything that may cause ignition.

   o Provide adequate ventilation.

☐ Smoking – Smoking is prohibited in the following areas:

   o All plant and office areas

   o Within 50 ft. of fuel equipment or storage tanks

   o **ALL** areas covered by company policy regarding smoking.

☐ Urns and Signs

   o Provide sand urns (or similar facilities) at the entrance to "No Smoking" areas.

   o Post "No Smoking" signs where smoking is prohibited. (The absence of such signs shall not be considered permission to smoke in a restricted area.)

☐ Open Flame Operations

   o Open flames of any kind are prohibited near flammable liquid storage or use. Do not carry matches or cigarette lighters in shirt pockets or any other upper pockets.

   o When any unusually hazardous operation is being conducted or when any unusual fire hazard exists, provide a fire guard with adequate fire fighting equipment.

☐ Combustible Metals

   o Use sharp tools when machining or fabricating magnesium. Do not use cutting fluids containing water.

   o Clean up dust or shavings immediately and place in a metal container for safe disposal.

o Have buckets of sand or a Class D fire extinguisher readily available in case of ignition.

☐ Other Safety Considerations

o Keep all oxygen lines and connections clean and free of oil and dirt to prevent violent, explosive reactions.

o Provide all motorized equipment with a muffler in good condition. Gasoline-powered equipment must have a carburetor air cleaner.

o Report all fire hazards to your supervisor or manager.

o Adhere to local regulations when they are more stringent.

HOUSEKEEPING RULES

Proper handling and disposal of rubbish is an integral part of the fire prevention process. Its success depends primarily on proper and regular disposal of combustible waste products. To maintain clean and orderly work areas, observe the following rules:

☐ Do not allow rubbish, rags, or waste material to accumulate.

☐ Provide unventilated metal containers with tight-fitting and self-closing covers for oily rags. Mark these containers: "For Oily Rags Only. Empty Daily."

☐ Place nothing in oily rag containers except cloth/paper items contaminated with oil or mild solvent.

☐ Empty waste containers daily. Dispose of all waste and rubbish in a safe manner and in accordance with local regulations.

☐ Do not allow ceiling tiles to remain removed except as required for temporary access above the ceiling for maintenance or installation of items above the ceiling.

ELECTRICAL EQUIPMENT

Observe the following rules in the operation and maintenance of electrical equipment:

☐ All electrical installations shall comply with the National Electrical Code and/or the appropriate local code.

☐ Where applicable, electrical equipment shall be of a type listed by the Underwriters Laboratories (UL) or Factory Mutual (FM) for the specific use and location designated.

☐ Electrical equipment and installations shall be maintained in a safe and serviceable condition.

☐ Repairs to electrical equipment may be made only by a qualified electrician or individual.

☐ Use only flashlights and drop cords approved by the Underwriters Laboratories for Hazardous Locations (Class I – Group D) around heavy concentrations of vapors or locations otherwise classified by the National Fire Protection Association (NFPA) as hazardous.

☐ Flexible cords must be in continuous lengths with no splices and must be 3-pronged (includes ground prong).

FLAMMABLE LIQUIDS

Flammable liquids are all liquids having flashpoints of 100°F or lower. Typical examples are acetone, alcohol, gasoline, methanol, thinners, etc. Observe the following safety requirements:

☐ Store flammable liquids in:

   o   Drums

   o   Unopened original containers

   o   Safety cans of a type listed by Factory Mutual or Underwriters Laboratories

   o   Approved flammable-storage cabinets labeled "Flammable – Keep Fire Away"

☐ Use only the quantity of flammable liquids for the minimum operational requirements. In other containers only the amount of flammable liquids that will be used during that shift may be stored on floor storage areas such as toolboxes.

☐ All drum spigots shall be FM or UL approved.

☐ A metallic bond shall be maintained between containers to prevent static spark ignition of vapors when flammable liquids are transferred. Use appropriate static bonding cables. Ground all dispensing drums.

☐ Unopened drums or containers of flammable liquids shall be stored vertically to minimize spillage and fire hazard.

RESPONSIBILITIES

☐ All employees should have a thorough knowledge of the use and location of all fire fighting equipment in their respective areas if they are expected to use such equipment.

☐ In the event of a fire, the person who first discovers the fire shall take immediate action as outlined in this section.

☐ They employer should ensure that each individual in the facility receives annual training in the use of fire extinguishers and identifying hazards involved with incipient-stage fire fighting. Hands-on training is not required. However, ensure that all employees know fire and evacuation procedures.

☐ The employer shall ensure that charts are posted showing the location of all fire fighting equipment, fire lanes, fire escape routes, and emergency shutoff valves and/or switches. Become familiar with the facility's type of sprinkler system.

☐ The employer shall ensure that all assigned fire fighting equipment is inspected.

☐ The employer shall determine the method of alarm and warning and when to use such warning for building evacuations. The employer shall notify all personnel through effective training of the types and sounds of the alarms or methods of notifying employees of an evacuation.

FIRE PROTECTION SYSTEMS

☐ Sprinkler Systems – Sprinklers provide automatic, 24-hour protection if properly maintained. The following rules shall be observed:

- o Do not place or pile material within 18 inches of sprinkler heads.

- o Mark with a 5-inch red stripe, a rectangular floor area 5 feet x 3 feet below and in front of each sprinkler control valve (may be a solid red block).

- o Mark the area "Keep Clear." Keep the area clear and unobstructed at all times.

- o Seal sprinkler control valves in the open position with a wire seal so that the valve cannot be closed without breaking the seal.

- o Inspect sprinkler control valves monthly. If a valve is found closed, investigate and determine the reason.

- o Plant supervisory personnel shall know the location of and how to operate the manual control valves to prevent unnecessary water damage.

> **NOTE:** Do not close a sprinkler control valve after a fire except upon authorization from the public fire official in charge.

- o If a sprinkler is shut down, arrange for manual operation until automatic operation is restored, and restrict all hot work from being performed in that area.

☐ Fire Doors

- o Fire door installations shall conform to National Fire Code requirements or applicable local regulations.

- o Fire doors may be self-closing or automatic, but must be marked "Fire Door, Keep Closed."

- o Self-closing doors shall be kept closed. **Do not block or obstruct them**.

- o Automatic closing doors are designed to close automatically in case of fire. Do not obstruct or block the doors. In the event of fire, immediately close the doors manually to eliminate drafts and to confine the fire. Do not wait for automatic operation.

INSPECTION, TESTING, AND MAINTENANCE OF FIRE PROTECTION SYSTEMS

☐ Guidelines for the inspection, testing, and maintenance of fire protection equipment are presented below. Where local and/or state authorities have more demanding requirements, they shall be met. Outside contractors shall be required to meet NFPA standards as well as local, state, or other applicable codes.

> **NOTE:** For fire extinguisher inspection testing and maintenance, refer to the Fire Extinguisher Section of this text.

- [ ] Monthly Test/Inspection
  - o All systems:
    - ◊ Record system pressure readings.
    - ◊ Verify that alarm devices are free from damage.
    - ◊ Ensure that there is no leakage from alarm drains or retard chambers.
    - ◊ Test deluge alarm sensor (deluge system only).
  - o Sprinkler control valves are:
    - ◊ In normal position (open)
    - ◊ Accessible
    - ◊ Not leaking
    - ◊ Properly identified.

- [ ] Annual Test/Inspection – Annual Test/Inspection must be conducted by a local contractor (or fire department) certified for fire equipment inspection in accordance with NFPA requirements and any other applicable local or state codes or regulations.

RECORDKEEPING

- [ ] Maintain a record of trained employees

- [ ] Training documentation

- [ ] Periodic inspections

---------- END OF SECTION ----------

Section 14
Flammable and Combustible Liquids

OVERVIEW

This section outlines the general requirements for the proper storage and handling of flammable and combustible liquids. However, the Flammable and Combustible Liquids Standard (1910.106) goes into great detail with regard to tank and vessel storage and the design of tanks and vessels. It also includes specific information for industrial plants where flammable and combustible liquids are incidental to the principal business (chemical plants, refineries, etc.), including service stations. These areas exceed the scope of this section, and the information herein will not be adequate if one of the aforementioned is the intended application.

It is important to know that state and/or local codes may supersede federal regulations if such codes are more restrictive than the federal regulations. Consideration must also be given to fire protection regulations (NFPA and/or local) and environmental (EPA, State, and/or local) requirements such as impact surveys and permitting for tanks.

REGULATORY COMPLIANCE

☐ §1910.106 Flammable and Combustible Liquids

☐ National Fire Protection Association (NFPA) 30 Flammable and Combustible Liquids Code 2003 Edition

☐ NFPA 251 Standard Methods of Tests of Fire Resistance of Building Construction and Materials 2006 Edition

TERMS & DEFINITIONS

☐ *Approved* – In compliance with, or listed by, a nationally recognized testing laboratory

☐ *Closed Container* – A container sealed by means of a lid or other device that neither liquid nor vapor will escape from it at ordinary temperatures.

☐ *Container* - Any can, barrel, or drum

☐ *Combustible Liquids (OSHA definition)* - Any liquid having a flashpoint at or above 100°F. Combustible liquids are divided into two classes.

   o **Class II:** Liquids with flashpoints at or above 100°F and below 140°F, except any mixture having components with flashpoints of 200°F or higher, the volume of which makes up 99% or more of the total volume of the mixture.

   o **Class III:** Liquids with flashpoints at or above 140°F. Subdivided into two subclasses:

◊ **Class IIIA:** Liquids with flashpoints at or above 140°F and below 200°F, except any mixture having components with flashpoints of 200°F or higher, the volume of which makes up 99% or more of the total volume of the mixture

◊ **Class IIIB:** Liquids with flashpoints at or above 200°F.

> **NOTE:** When a combustible liquid is heated for use to within 30°F of its flashpoint, it shall be handled in accordance with the requirements for the next lower class of liquids.

☐ *__Fire Area__* – An area of a building separated from the remainder of the building by construction having a fire resistance of at least 1-hour and having all communicating openings (doors, etc.) properly protected by an assembly having a fire-resistance rating of at least 1 hour.

☐ *__Flammable Liquid (OSHA definition)__* – Any liquid having a flashpoint below 100°F. These liquids shall be known as Class I liquids and are divided into three subclasses as follows:

o **Class IA:** Liquids with flashpoints below 73°F and a boiling point below 100°F.

o **Class IB:** Liquids with flashpoints below 73°F and a boiling point at or above 100°F.

o **Class IC:** Liquids with flashpoints at or above 73°F and below 100°F.

☐ *__Flashpoint__* - Minimum temperature at which a liquid gives off enough vapor in sufficient concentration to form an ignitable mixture with air near the surface of the liquid.

☐ *__Portable Tank__* - A closed container having a liquid capacity over 60 gallons (US) and not intended for fixed installation.

☐ *__Safety Can__* - An *approved* container of not more than 5 gallons capacity, having a spring-closing lid and spout cover, and so designed that it will safely relieve internal pressure when subjected to fire exposure. Modification to such a container is not allowed.

☐ *__Ventilation__* - As specified in this section, ventilation involves the prevention of fire and explosion. It is considered adequate to prevent accumulation of significant quantities of vapor-air mixtures in concentration of over 1/4 of the lower flammable limit.

FIRE PREVENTION

BP: Refer to the MSDS for more information about the products you work with, including the product's ingredients, potential health effects, hazards, emergency procedures, storage, disposal information, and what personal protective equipment should be worn.

☐ Extinguishers: Either fire hoses or portable fire extinguishers shall be available at locations where flammable or combustible liquids are stored.

o Any room used for storing these liquids shall have at least one portable fire extinguisher located just outside the door to the room. The extinguisher will be located within 10 feet of the door and will have a rating of at least 12-B.

o Any Class I or Class II liquid storage area located outside a storage room but inside a building shall have at least one portable fire extinguisher with a rating of 12-B or more located not under 10 feet or over 25 feet from the door.

☐ Open Flames, Smoking: Open flames and smoking shall not be permitted in flammable or combustible liquid storage areas.

☐ Water-Reactive Materials: Materials that will react with water shall not be stored in the same room with flammable or combustible liquids.

SOURCES OF IGNITION

☐ Adequate precautions shall be taken to prevent ignition of flammable vapors. Sources of ignition include open flames, hot surfaces, frictional heat, static, electrical and mechanical sparks, spontaneous ignition, including heat-producing chemical reaction and radiant heat.

☐ Bonding and Grounding: Class I liquids shall not be dispensed into containers unless the nozzle and the container are electrically interconnected, and the dispensing unit is electrically connected (grounded) to either building steel or earth.

STORAGE REQUIREMENTS

☐ Quantity Limitations: Flammable or combustible liquids shall be stored in tanks or closed containers with provisions of this section except as noted below:

  o The quantity of liquid that may be located outside an inside storage room or storage cabinet in a building or in any one fire area of a building shall not exceed:

    ◊ 25 gallons of Class IA liquids in containers

    ◊ 120 gallons of Class IB, IC, II, or III liquids in containers, or

    ◊ 660 gallons of Class IB, IC, II, or III in a single portable tank

☐ Handling at Point of Final Use: Flammable or combustible liquids shall be stored in tanks or closed containers with provisions of this section except as noted below:

  o Flammable liquids shall be kept in covered containers when not actually in use.

  o Where flammable or combustible liquids are used or handled, except in closed containers, means shall be provided to dispose promptly and safely of leakage or spills.

  o Class I liquids may be used only where there are no open flames or other sources of ignition within the possible path of vapor travel.

☐ Transfer of Liquids:

  o From Containers: Areas in which flammable or combustible liquids are transferred from one container to another shall be separated from other operations in the building by an adequate distance or by construction having adequate fire resistance.

  o Within a Building: Flammable or combustible liquids shall be drawn from or transferred into vessels, containers, or portable tanks within a building **only** through one of the following:

    ◊ A closed piping system

    ◊ From safety cans

    ◊ By means of a device drawing through the top or from a container

    ◊ Portable tank by gravity through an approved self-closing valve. Transferring by means of air pressure on the container or portable tanks is prohibited

DESIGN, CAPACITY, AND CONSTRUCTION OF CONTAINERS

☐ Only approved containers and portable tanks shall be used.

☐ Emergency Venting: Each portable tank shall be provided with one or more devices installed in the top with sufficient emergency venting capacity to limit internal pressure under fire exposure conditions to 10 psig or 30% of the bursting pressure of the tank, whichever is greater.

- o At least one pressure-actuated vent having a minimum capacity of 6,000 cubic feet of free air shall be used.

- o The vent shall be set to open at not less than 5 psig.

- o If fusible vents are used, they shall be actuated by elements that operate at a temperature not exceeding 300°F.

☐ Size of Containers: Capacities of flammable and combustible liquid containers shall be in accordance with **Table 1**, except that glass or plastic containers of no more than 1-gallon capacity may be used for a Class IA or IB flammable liquid if such liquid would corrode a metal container, be rendered unfit for its intended use by contact with metal, or excessively corrode a metal container so as to create a leakage hazard.

Table 1: Flammable/Combustible Liquids

Container Type	Class IA	Class IB	Class IC	Class II	Class III
Glass or Approved Plastic Gal.	I pt.	I qt.	I gal.	I gal.	I gal.
Metal (other than DOT drums) Gal.	I gal.	5 gal.	5 gal.	5 gal.	5 gal.
Safety Cans	2 gal.	5 gal.	5 gal.	5 gal.	5 gal.
Metal Drums (DOT specs.)	60 gal.	60 gal.	60 gal.	60 gal.	60 gal.
Approved Portable Tanks	660 gal.	660 gal.	660 gal.	660 gal.	660 gal.

DESIGN, CONSTRUCTION, AND CAPACITY OF STORAGE CABINETS

☐ Capacity Limits:

- o Not more than 120 gallons of Class I, Class II, and Class IIIA liquids shall be stored in a storage cabinet.

- o Of this total, not more than 60 gallons shall be of Class I and Class II liquids.

- o Not more than three such cabinets shall be located in a single fire area, except that in an industrial occupancy additional cabinets shall be permitted to be located in the same fire area if the additional cabinet or group of not more than three cabinets is separated from other cabinets or groups of cabinets by at least 100 feet.

☐ Fire Resistance: Storage cabinets shall be designed and constructed to limit the internal temperature to not more than 325°F when subjected to a 10-minute fire test (see NFPA 251). All joints shall remain tight, and the door shall remain securely closed during the test. Cabinets shall be labeled in conspicuous lettering "Flammable – Keep Fire Away."

☐ Cabinet Construction: The bottom, top, door, and sides should be at least 18-gauge sheet iron and double-walled with 1½-inch air space. Joints shall be riveted, welded, or tightened by some equally effective means. The door shall be provided with a three-point lock and the doorsill shall be raised at least 2 inches above the bottom of the cabinet.

☐ Cabinets:

- o Cabinets positioned side-by-side shall have a minimum of 6 inches of clear space between each other.

- o The tops of cabinets shall not be used for storage.

- o Cabinets that become damaged, altered, or otherwise defective shall be considered no longer serviceable for use.

o Cabinets shall have the vent bung installed in all cases except when the cabinet is vented through the bung to the outside of the building.

BP: For flammable and combustible liquid storage, flammable liquids storage cabinets are the most frequently used (and often misused) methods throughout industry. There are a few things to keep in mind when planning storage and use of flammable storage cabinets:

◊ Use flammable liquids storage cabinets for flammable and combustible liquids storage only. Flammable liquids storage cabinets can be very expensive and an inappropriate use of space for products such as oil or water-based paints.

◊ Do not store combustible materials in flammable liquids storage cabinets (e.g., cardboard boxes, cloth materials, etc.).

◊ Do not store materials on top of cabinets. To prevent this from occurring, there are cabinet "toppers" that can be purchased or fabricated in a metal shop that fit on top of cabinets to discourage placing other items on top (because of their "teepee" shape).

◊ Never use a flammable storage cabinet for personal items such as respirators or food products.

STORAGE IN BUILDINGS

☐ The total quantity of liquids within a building will not be restricted, but the arrangement of storage will comply with **Table 2**:

Table 2: Indoor Container Storage
(Numbers in parentheses indicate corresponding number of 55-gallon drums.)

Class Liquid	Storage Level	Protected storage maximum per pile		Unprotected storage maximum per pile	
		Gallons	**Height**	**Gallons**	**Height**
IA	Ground and Upper Floors	2,750 (50)	3 ft. (1)	660 (12)	3 ft. (1)
IB	Ground and Upper Floors	5,500 (100)	6 ft. (2)	1,375 (25)	3 ft. (1)
IC	Ground and Upper Floors	16,500 (300)	6 ft. (2)	4,125 (75)	3 ft. (1)
II	Ground and Upper Floors	16,500 (300)	9 ft. (3)	4,125 (75)	9 ft. (3)
III	Ground and Upper Floors	55,000 (1,000)	15 ft. (5)	13,750 (250)	12 ft. (4)

Note 1: *When two or more classes of materials are stored in a single pile, the maximum quantity permitted in that pile shall be equal to that of the smallest class.*

Note 2: *Aisles shall be provided so that no container is more than 12 feet from an aisle. Main aisles shall be at least 8 feet wide and side aisles at least 4 feet wide.*

Note 3: *Each pile shall be separated from each other by at least 4 feet.*

Note 4: *No pile shall be closer than 3 feet to the nearest beam, chord, girder, or other obstruction, and shall be 3 feet below sprinklers or other overhead fire protection systems.*

---------- END OF SECTION ----------

Section 15
Guarding Floor and Wall Openings

OVERVIEW

This section provides the requirements for the safe guarding of wall openings, floor openings, and holes.

REGULATORY COMPLIANCE

☐ §1910.23 Guarding Floor and Wall Openings and Holes

TERMS AND DEFINITIONS

☐ ***Floor Hole*** – An opening measuring less than 12 inches but more than 1 inch in its least dimension, in any floor, platform, pavement, or yard, through which materials but not persons may fall; such as a belt hole, pipe opening, or slot opening.

☐ ***Floor Opening*** – An opening measuring 12 inches or more in its least dimension, in any floor, platform, pavement, or yard through which persons may fall; such as a hatchway, stair or ladder opening, pit, or large manhole. Floor openings occupied by elevators, dumb waiters, conveyors, machinery, or containers are excluded from this subpart.

☐ ***Handrail*** – A single bar or pipe supported on brackets from a wall or partition, as on a stairway or ramp, to furnish persons with a handhold in case of tripping.

☐ ***Platform*** – A working space for persons, elevated above the surrounding floor or ground; such as a balcony or platform for the operation of machinery and equipment.

☐ ***Runway*** – A passageway for persons, elevated above the surrounding floor or ground level, such as a foot-walk along shafting or a walkway between buildings.

☐ ***Standard Railing*** – A vertical barrier erected along exposed edges of a floor opening, wall opening, ramp, platform, or runway to prevent falls of persons.

☐ ***Standard Strength and Construction*** – Any construction of railings, covers, or other guards that meets the requirements of §1910.23 Guarding Floor and Wall Openings and Holes.

☐ ***Stair Railing*** – A vertical barrier erected along exposed sides of a stairway to prevent falls of persons.

☐ ***Toeboard*** – A vertical barrier at floor level erected along exposed edges of a floor opening, wall opening, platform, runway, or ramp to prevent falls of materials.

☐ ***Wall Hole*** – An opening less than 30 inches but more than 1 inch high, of unrestricted width, in any wall or partition; such as a ventilation hole or drainage scupper.

☐ ***Wall Opening*** – An opening at least 30 inches high and 18 inches wide, in any wall or partition, through which persons may fall; such as a yard-arm doorway or chute opening.

PROTECTION FOR FLOOR OPENINGS

☐ Every stairway opening shall be guarded by a standard railing system.

☐ Every ladder-way floor opening or platform shall be guarded by a standard railing system.

☐ Every hatchway and chute floor opening shall be guarded by one of the following:

    o Hinged floor opening cover with standard railings. When not in use, the opening shall be guarded at both top and intermediate positions by removable standard railings.

    o Removable railing with toeboard on not more than two sides of the opening and fixed standard railings with toeboards on all other sides.

☐ Every skylight floor opening and hole shall be guarded by a standard skylight screen or fixed standard railing system.

☐ A floor opening cover shall guard every pit or trap door floor opening (when infrequently used). All sides shall be protected by a removable standard railing system when the cover is not in place (or someone must constantly attend the opening).

☐ A manhole cover shall guard every manhole floor opening. All sides shall be protected by a removable standard railing system when the cover is not in place (or someone must constantly attend the opening).

☐ Every temporary floor opening shall be protected by a standard railing system (or someone must constantly attend the opening).

☐ Every floor hole into which persons can accidentally walk shall be guarded by:

    o Standard railing system on all exposed sides.

    o <u>Floor Hole Cover</u> – All sides shall be protected by a removable standard railing system when the cover is not in place (or someone must constantly attend the opening).

    o A cover that leaves no openings more than 1-inch wide shall protect every floor hole into which persons cannot accidentally walk (on account of fixed machinery, equipment, or walls). The cover shall be securely held in place to prevent tools or materials from falling through.

☐ Where doors or gates open directly on a stairway, a platform shall be provided, and the swing of the door shall not reduce the effective width to less than 20 inches.

PROTECTION FOR WALL OPENINGS AND HOLES

☐ Wall openings that have a drop of 4 feet or more shall be guarded by one of the following:

    o Rails

    o Rollers

    o Picket fence

    o Half door

    o Any other equivalent barrier to prevent falling

☐ Extension platforms on which materials can be hoisted shall have side rails or equivalent guards.

☐ Every chute wall opening where there is a drop of more than 4 feet shall be guarded just as any other wall opening.

☐ Every window wall opening from which there is a drop of 4 feet and where the bottom of the opening is less than 3 feet above the platform or landing must also be guarded.

☐ Every temporary wall opening must also be guarded.

☐ Where there is a hazard of materials falling through a wall hole, the opening must be guarded by a 4-inch toeboard or a screen.

PROTECTION OF OPEN-SIDED FLOORS, PLATFORMS, AND RUNWAYS

☐ Open-sided floors or platforms that are 4 feet or more off the adjacent floor shall be guarded by a standard railing system (except for entrances to a ramp, stairs, or fixed ladder).

☐ A railing system shall be provided on elevated platforms when:

   o People can pass below

   o Moving machinery is above

   o Potential hazard of falling material is present

☐ Runways 4 feet or more above the surface below shall be guarded by a standard railing system.

☐ Toeboards shall be provided on runways where there is a potential hazard of falling materials, parts, or tools.

STAIRWAY RAILINGS AND GUARDS

☐ Every flight of stairs having four or more risers shall be equipped with standard handrails.

☐ On stairways 44 inches wide or less with both sides enclosed, the handrail should be on the right-hand side.

☐ On stairways 44 inches wide or less with one side open, the handrail should be on the open side.

☐ On stairways 44 inches wide or less with both sides open, handrails must be provided on both sides.

☐ On stairways more than 44 inches but less than 88 inches wide, a handrail is required on the enclosed side as well as each open side.

☐ On stairways 88 or more inches wide, one handrail on each enclosed side, one stair railing on each open side, and one stair railing approximately in the middle.

☐ Winding stairs must be equipped with a handrail offset to prevent the person from walking on all treads less than 6 inches wide.

RAILING, TOEBOARDS, AND COVERS

☐ A standard railing shall consist of a top rail, intermediate rail, toeboards, and posts (except at the opening).

☐ All railings shall have a vertical height of 42 inches from the floor to the upper surface of the top rail.

☐ The top rail shall be smooth and free from jagged edges.

☐ The intermediate rail shall be approximately halfway (21 inches from the floor) between the floor and the top rail.

☐ The ends of the rail shall not overhang where a possible projection hazard can occur.

☐ A standard toeboard shall be at least 4 inches from the top of the toeboard to the platform floor.

☐ The toeboard shall be securely in place and no more than 1/4-inch from the bottom of the toeboard to the platform floor.

☐ All railings shall withstand 200 lbs. of pressure applied in any direction.

FLOOR OPENING COVERS

☐ Trench or conduit covers and their supports must be able to withstand 20,000 pounds when placed in roadways.

☐ Floor opening covers shall not project from the ground level more than 1 inch.

☐ The edges of the floor opening cover shall be beveled at 30 degrees or less.

☐ Skylight screens should support 200 pounds applied perpendicularly at any one area on the screen.

☐ Wall opening barriers such as rails, rollers, pickets, and half doors must be capable of supporting 200 pounds applied in any direction except up.

☐ Grab handles shall be placed vertically on each side of the wall opening. The center of each handle shall be approximately 4 feet from the ground.

☐ Grab handles shall be not less than 12 inches in length and at least 3 to 4 inches between the handle and the wall.

☐ Grab handles must be capable of supporting 200 pounds applied in any direction on the handle.

----------END OF SECTION----------

Section 16
Hand and Portable Power Tools

OVERVIEW

The purpose of this section is to establish general safe work practices involving hand and portable powered tools. This includes the use of small equipment such as mowers and jacks. Use of safe work practice and personal protective equipment is essential. Tools and equipment should always be inspected prior to each use and removed from service if defective or in questionable condition.

REGULATORY COMPLIANCE

☐ §1910.242 Hand and Portable Power Tools and Equipment, General

☐ §1910.243 Guarding of Portable Powered Tools

☐ §1910.244 Other Portable Tools and Equipment

GUARDING REQUIREMENTS

☐ Portable Circular Saws – Portable circular saws with a diameter greater than 2 inches must be equipped with guards above and below the base plate or shoe.

   o **The upper guard** must cover the blade teeth, except for the minimum arc needed for the base to be tilted for bevel cuts.

   o **The lower guard** must cover the blade teeth, except for the minimum arc needed for extraction and contact. When the tool is withdrawn from the work, the lower guard should automatically return to the covering position.

☐ Constant pressure switches or controls that shut off power when pressure is released are required on:

   o Percussion tools without positive accessory holding means: electric, hydraulic, or pneumatic chain saws

   o All hand-held powered tools, including:

      ◊ Circular saws with blade diameters greater than 2 inches

      ◊ Drills

      ◊ Tapers

      ◊ Fastener drivers

      ◊ Horizontal, vertical, and angle grinders with wheels greater than 2 inches in diameter

      ◊ Disc sanders with discs greater than 2 inches in diameter

◊ Belt sanders

◊ Reciprocating saws

◊ Saber, scroll, and jigsaws with blade shanks greater than 1/4-inch

☐ Portable belt sanding machines shall have guards at each nip point where the sanding belt runs onto a pulley to effectively prevent operator hands or fingers from coming in contact with the nip points. The unused run of the sanding belt must be guarded against accidental contact.

☐ Portable Abrasive Wheels

o Only machines with safety guards should use abrasive wheels, unless the wheels are

◊ Being grounded in internal work

◊ Mounted and 2 inches or smaller in diameter

o A safety guard shall cover the spindle end, nut, and flange projections. The safety guard shall be properly aligned with the wheel. The strength of the fastenings must exceed the strength of the guard.

o The spindle end, nut, and outer flange may be exposed if the safety guards on that particular operation protect the operator.

o Side covers of the wheels may be omitted if the work entirely covers the side of the wheel.

o The spindle end, nut, and outer flange may be exposed on portable machines with cutoff wheels.

☐ Cup Wheels – Cup wheels must be protected by:

o Safety guards

o Special revolving cup guards that mount behind the wheel and turn with it. Clearance between the wheel side and the guard shall not exceed 1/16 inch.

o Safety guards on right-angle head or vertical portable grinders shall have a maximum exposure angle of 180°, with the guard located between the operator and wheel during use. The guard shall deflect pieces of an accidentally broken wheel away from the operator.

o The maximum angular exposure of the grinding wheel periphery and sides for safety guards used on other portable grinding machines shall not exceed 180°, and the top half of the wheel shall be enclosed.

MOUNTING OF WHEELS

Exception: Natural sandstone wheels and metal, wooden, cloth, or paper discs having a layer of abrasive on the surface are excluded from these mounting and inspection requirements

☐ Immediately before mounting the wheel, the operator shall inspect and sound test all wheels to ensure they have not been damaged in transit, storage, or otherwise.

> **NOTE:** When tapped gently with a metallic object, an intact wheel will ring; a damaged one will not.

☐ To avoid excessive pressure from mounting and spindle expansion, a controlled spindle, wheel sleeves, or adapters shall be used. The machine spindle shall be nominal (standard) size plus a negative .002 inch (–.002 inch). The wheel hole will be oversized for safety clearance under operating heat and pressure.

☐ All contact surfaces of wheels, blotters, and flanges shall be flat, clean, and clear.

☐ A bushing in the wheel hole shall not exceed the width of the wheel and shall not contact the flanges.

☐ The operator will also check the spindle speed of the machine before mounting the wheel to ensure it does not exceed the maximum operating speed on the wheel.

SHUTOFF CONTROLS – REQUIREMENTS

☐ A lock-on control for turning off/on by a single motion of the finger(s) can be used.

☐ Constant pressure throttle controls that shut off power when the pressure is released must be on all hand-held gasoline-powered chain saws.

☐ Positive on/off controls or controls described above must be on all other hand-held powered tolls, including but not limited to:

   o Platen sanders

   o Grinders with wheels 2 inches in diameter or less

   o Disc sanders with discs 2 inches in diameter or less

   o Routers

   o Planers

   o Laminate trimmers

   o Nibblers

   o Shears

   o Saber, scroll, and jigsaws with blade shanks 1/4-inch wide or less

PNEUMATIC POWERED TOOLS AND HOSES

☐ A tool retainer must be installed on each piece of equipment used.

☐ Hoses and hose connections that conduct compressed air to equipment must be designed for the appropriate pressure and service to which they are subjected.

GROUNDING OF PARTS

☐ Non-current-carrying metal parts of the following types of (cord and plug-connected) equipment that may become energized must be grounded:

   o Tools used in hazardous-classified locations (flammable atmospheres).

   o Tools operated at over 150 volts.

   o Equipment used with water or wet processes, such as sump pumps and washing machines.

   o Hand held motor-operated tools.

   o Portable and mobile X-ray equipment.

   o Portable hand lamps.

   o Tools likely to be used in wet locations or by employees standing on the ground or on metal.

JACKS

☐ Loading and Marking

  o The operator will ensure that the jack used has a rating sufficient to lift and sustain the load.

  o The rated load must be legibly, permanently, and prominently marked on the jack by casting, stamping, or other suitable means.

☐ Operation and Maintenance

  o The base of the jack shall be blocked if there is no firm foundation.

  o A block shall be placed between the cap and the load if there is a possibility the cap will slip.

  o The limit of travel shall be determined by watching the stop indicator, which shall be kept clean. Limit overruns are not permitted.

  o Hydraulic jacks exposed to freezing temperatures must be supplied with antifreeze.

  o After the load has been raised, it must immediately be cribbed, blocked, or otherwise secured.

  o Jacks must be properly lubricated.

☐ Inspections

  o Each jack must be thoroughly inspected according to service conditions.

  o For constant or intermittent use at a single location, inspect once every 6 months.

  o For jacks sent off for special work, inspect when sent out and upon return.

  o For jacks subjected to abnormal load/shock, inspect immediately before and after usage.

  o Out-of-order jacks must be tagged accordingly and not used until repaired.

  o Repair or replacement parts must be examined for possible defects.

POWERED LAWNMOWERS (INDUSTRIAL USE)

☐ All mowers are required to meet American National Standards Institute (ANSI) guidelines.

☐ Walk-behind and riding mowers shall be installed with a deadman's control switch which will cut off the motor when pressure is released.

☐ Blades on mowers shall have guards installed with a discharge opening.

☐ Discharge from mowers shall always be directed away from the user.

☐ Power riding mowers are required to have a stop installed to prevent the steering wheel from turning to a point that could cause jack-knifing or locking.

---------- END OF SECTION ----------

Section 17
Hazard Communication Program

OVERVIEW

The purpose of this section and the Hazard Communication Program is to provide information to employees about the hazards of the chemicals and products used in their workplace. Included in this program is information for the use of Material Safety Data Sheets (MSDSs), chemical product labeling, handling and storage, training, documentation, and record keeping requirements.

The Hazard Communication (HazCom) Program shall be designed to ensure that all employees and contracted personnel are made aware of and specifically trained in the hazards associated with any hazardous chemical they may be exposed to while working on the employer's site.

REGULATORY COMPLIANCE

☐ §1910.1200 – Hazard Communication

TERMS/DEFINITIONS

☐ **_HMIS_** - refers to the Hazardous Materials Identification System. This system is used by AIH to identify hazards associated with chemicals in the workplace that are not already labeled with hazard warning data. Corresponding codes may be found in the Hazard Index subsection.

☐ **_MSDS_** - Material Safety Data Sheet(s). Data sheets that contain safety information about chemical products.

TRAINING/QUALIFICATIONS

☐ All employees who are or may be exposed to hazardous and/or toxic chemicals while performing duties in their work area must receive training on these hazards at the time of their initial assignment and whenever a new hazard is introduced into their work area. Persons who have not received this training shall not be allowed to work in areas where hazardous chemicals or substances exist. Employees must receive recurrent training within 12 months of initial training and at 12-month intervals thereafter.

☐ Each employee must be informed of the following:
   o   The requirements of the Hazard Communication Program and OSHA regulations.
   o   Any operation in the employee's work area where hazardous chemicals are present.
   o   The location and availability of the written Hazard Communication Program.
   o   The location and availability of the workplace chemical list and MSDSs.

☐ Content must include the following:

- o Methods to observe the work area to detect the presence or release of a hazardous chemical by visual appearance or the odor of the chemical when being released.

- o Training on the physical and health hazards of the chemicals in the work area.

- o Training on "Right-to-Know" of hazards (how to read and understand MSDSs).

- o An explanation of the company's labeling system.

- o A quiz. The trainer should review any incorrect answers from the quiz to assure the employee has an understanding of the material and its application.

OUTSIDE CONTRACTOR NOTIFICATION

☐ To ensure that any outside contractors work safely at the worksite, contractors must be provided with the following information:

- o Access to or copies of the Hazard Communication Program as well as the Workplace Chemical List and MSDSs.

- o Information and location for any hazardous substances that they may be exposed to while working on the job site.

BP: Contractors should be required to review this information with their employees prior to the start of work. In addition, request that contractors disclose information regarding chemical products that will be brought onsite. It is the employer's responsibility to evaluate these to determine if they should be allowed onsite and around employees.

MATERIAL SAFETY DATA SHEETS

☐ A Material Safety Data Sheet (MSDS) is a document that describes the physical and chemical properties of products, their physical and health hazards, and precautions for safe handling and use.

☐ The manufacturers and distributors of chemical products furnish MSDSs. A MSDS must be maintained for each chemical used at the location.

BP: In cases where a department receives repetitive shipments of the same chemical from a supplier, the MSDS may already be on file. Check the revision date and the date of the MSDS. If they are the same, discard the copy. If more recent, keep a copy on hand and send the original to the individual or department responsible for maintaining MSDSs.

☐ MSDSs shall available for viewing by employees on any shift. This can be accomplished by providing them via hard copy, electronically, or fax-on-demand as long as the method used can be proven reliable or backed up by a secondary means.

☐ MSDS Reference Explanation
See the sample MSDS for the corresponding letter reference explanation:

A) **Section I**
Chemical Identification – The identity of the material must indicate specific chemical name, common name, and trade name, if applicable. The specific chemical name is one that will permit any qualified chemist to identify the structure of the compound(s) present to enable further searching of literature for desired information.

B) Name and address of the chemical manufacturer. Phone number to be used for information or an emergency.

C) **Section II**
Hazardous Ingredients/Identity Information – This section lists what's in the substance that may cause harm. The chemical name(s) of the substance, as well as the common name of the substance must be indicated. It also lists the concentration of the chemicals to which you can safely be exposed to within and 8-hour period (PEL – Permissible exposure limit, TLV – Threshold limit value).

D) **Section III**
Physical/Chemical Characteristics – This section describes the substance's appearance, odor, and other chemical characteristics. These are things that can affect the degree of hazard you face in different work situations:
- Boiling Point/Melting Point – to help prevent a potentially dangerous change in state, as from a liquid to a breathable gas.
- Vapor Pressure, Density, and Evaporation Rate – important for flammable or toxic gases and vapors.

E) Solubility in Water and Specific Gravity – identifies if a chemical will dissolve in water, sink, or float.

F) Normal Appearance and Odor – to help recognize anything different or possibly dangerous.

G) **Section IV**
Flash Point – Indicates the minimum temperature at which a liquid gives off a vapor in sufficient concentration to ignite.

H) Flammable Limits – Indicates the concentration of the substance, in the form of a gas or a vapor, that is required for it to ignite.

I) LEL and UEL – Lowest and highest (respectively) concentration (percentage of substance in the air) that will produce a flash of fire when an ignition source is present. At higher concentrations, the mixture is too "rich" to burn.

J) This section describes methods of fire fighting including proper extinguishing media (i.e. ABC, CO_2, foam, etc.). This section also describes unusual fire and explosion hazards.

K) **Section V**
Reactivity Data – This section indicates whether the substance is stable or unstable and conditions to avoid such as direct sunlight or any condition which may cause a dangerous reaction.

L) Incompatibility – Lists materials, chemicals, and other substances to avoid which may cause the chemical to burn, explode, or release dangerous gases.

M) **Section VI**
Health Hazard Data – This is probably the most important section of the MSDS. It describes how a substance can enter your body (e.g. inhalation [respired], through the skin [absorbed], or swallowing [consumed]).

N) Health Hazards (Acute and Chronic) – This section lists the specific possible health hazards which could result from exposure, both acute and chronic.

- Some effects like minor skin burns, are acute (they show up immediately after exposure). Others, like liver damage, are chronic (they are often the result of exposure long ago or repeated small exposures over a long period of time).

- If the substance or chemical is considered or believed to be a carcinogen, it will be indicated here.

O) This section lists the signs and symptoms of overexposure such as eye irritation, nausea, dizziness, skin rashes, headache, and burns.

P) If exposure to the substance could aggravate an existing medical condition, this information will be indicated here.

Q) This section describes emergency and first aid procedures to be followed in case of overexposure until medical help arrives. If these instructions are unclear, seek medical assistance immediately.

R) **Section VII**
Precautions for Safe Handling and Use – Explains procedures for spills, leaks, or any release and includes how to handle and store the substance, how to safely dispose of the substance, and other precautions.

S) This section describes the proper procedures to be followed in the case of spills or leaks, and the equipment needed.

NOTE: Not all MSDSs will indicate the personal protective equipment recommended when cleaning up a spill. Always reference section VIII – Control Measures, for specific types of recommended personal protective equipment to be worn prior to cleaning up a spill. Always notify your supervisor when a chemical spill or leak occurs, no matter what chemical, no matter how small.

T) This section lists the proper disposal procedures.

U) This section describes the proper handling and storage procedures, as well as other precautions to be followed (e.g. grounding containers during transfer of flammable materials to prevent static electricity as an ignition source).

V) **Section VIII**
Control Measures – Control Measures for preventing or reducing the chance of harmful exposure are listed in this section including personal protective equipment to be worn, work and hygiene practices, and ventilation requirements.

W) Proper respiratory equipment and ventilation information.

X) Gloves, eye protection, and other personal protective clothing recommended or required. List in this section work or hygiene practice which should be exercised when handling this substance or chemical such as taking a shower, washing work clothes, destroying soiled clothing, washing hands, etc.

NOTE: The HazCom Standard does not describe the format to be used in completing the MSDS. However, the standard does specify the information the MSDS must contain (Title 29 CFR 1910.1200(g)). Many manufacturers will develop their own MSDS in a format of their desire, but must meet the HazCom Standard's requirements. The OSHA Form 174 meets the HazCom Standard's requirements. An example of this form (Figure 1) follows:

Material Safety Data Sheet

May be used to comply with
OSHA's Hazard Communication Standard,
29 CFR 1910.1200. Standard must be
consulted for specific requirements.

U.S. Department of Labor

Occupational Safety and Health
(Non-Mandatory Form)
Form Approved
OMB No. 1218-0072

IDENTITY (As Used on Label and List) Isopropyl Alcohol (IP) **A**	Note: Blank spaces are not permitted. If any is not applicable, or no information is available, the space must be marked to indicate that.

Section I

Manufacturer's Name **B** Address (Number, Street, City, State, and ZIP Code)	Emergency Telephone Number	(800) 274-5263
Ashland Chemical Company	Telephone Number for Information	(614) 889-3333
P.O. Box 2219	Date Prepared	3/1/1994
Columbus, OH 43216	Signature of Preparer	Howard Yugo

Section II --- Hazardous Ingredients/Identity Information

Hazardous Components (Specific Chemical Identity; Common Name(s)) **C**	OSHA PEL	ACGIH TLV	Other Limits Recommended	% (optional)
Isopropanol CAS #: 67-63-0	400 ppm	400 ppm	500 ppm STEL	99

Section III --- Physical/Chemical Characteristics

Boiling Point **D**	180 F	Specific Gravity (H2O = 1)	0.789
Vapor Pressure (mm Hg.)	33	Melting Point	N/A
Vapor Density (AIR = 1)	2.07	Evaporation Rate (Butyl Acetate = 1)	7.7
Solubility in Water	N/A		
Appearance and Odor **E**		Clear with solvent odor	

Section IV --- Fire and Explosion Hazard Data

Flash Point (Method Used) (TCC) **F**	53 F	Flammable Limits **G**	LEL **H** 2.0%	UEL 12.0%
Extinguishing Media **I**		Alcohol foam or carbon dioxide or dry chemical		
Special Fire Fighting Procedures		Use SCBA w/full facepiece operated in pos. press. demand mode		
Unusual Fire and Explosion Hazards		Never use welding or cutting torch near drum (even empty) Ground and bond when material is transferred between containers		

(Reproduce locally)

OSHA 174, Sept. 1985

Section V --- Reactivity Data **J**

Stability	Unstable		Conditions to Avoid	Don't use with Aluminum equip. at temps above 120 F
	Stable	X		

Incompatibility (Materials to Avoid) **K** Strong oxidizing agents, acids, chlorine, acetaldehyde, ethylene oxide, isocyanate

Hazardous Decomposition or Byproducts

Hazardous Polymerization	May Occur		Conditions to Avoid
	Will Not Occur	· X	

Section VI --- Health Hazard Data **L**

Route(s) of Entry:	Inhalation?	X	Skin?	X	Ingestion?	X	

Health Hazards (Acute and Chronic) **M** Material has been shown to cause harm to the fetus in lab animal studies

Harm to the fetus occurs only at exposure temps that harm the pregnant animal. Relevance to humans uncertain

Carcinogenicity: **N**	NTP? No	IARC Monographs? No	OSHA Regulated? Not as carcinogen

Not found to be carcinogenic

Signs and Symptoms of Exposure **O** Gastrointestinal irritation (nausea, vomiting, diarrhea), impaired coordination

CNS depression (dizziness, fatigue, headache, etc.), low blood pressure, respiratory depression, coma

Medical Conditions Generally Aggravated by Exposure **P** Overexposure has been found to cause the following effects in laboratory

animals: mild, reversible liver effects

Emergency and First Aid Procedures **Q** Skin, remove contaminated clothing, wash with soap & water; Eyes, flush

for at least 15 min.; Swallowed, induce vomiting if conscious; Breathed, move away from exp., apply CPR if nec

Section VII --- Precautions for Safe Handling and Use **R**

Steps to Be Taken in Case Material Is Released or Spilled **S** Small spill, absorb on vermiculite, floor absorbent, or other

material. Large spill, eliminate ignition sources, persons not wearing PPE should be excluded from the area, stop

spill source at once, prevent from entering drains, if runoff occurs, notify authorities as required

Waste Disposal Method **T** Dispose of in accordance with all applicable local, state, and federal regs.

Precautions to Be Taken in Handling and Storing **U** Store away from ignition sources, label accordingly

Other Precautions Containers of material may be hazardous when empty, follow hazard precautions

given in this data sheet

Section VIII --- Control Measures **V**

Respiratory Protection (Specify Type) **W** If PEL/TLV exceeded, must wear NIOSH/MSHA app. air supp or neg press.

Ventilation	Local Exhaust Sufficient mechanical	Special	
	Mechanical (General)	Other	

Protective Gloves	X Chem. Resistant	Eye Protection	Safety glasses (Goggles recommended)

Other Protective Clothing or Equipment To prevent repeated skin contact, wear impervious clothing and boots

Work/Hygienic Practices

Figure 1

WORKPLACE CHEMICAL LIST

☐ The Workplace Chemical List identifies materials and chemical products that are or contain ingredients classified by OSHA and/or the DOT as potentially hazardous to personnel. The Hazard Communication Standard requires that a list of hazardous chemicals be included as part of the employer's Hazard Communication Program. The list may also serve as an inventory for which an MSDS is maintained on file.

SAFETY AND PIPING COLOR CODES

☐ OSHA Safety Colors: *The Occupational Safety and Health Act (OSHA) requires that all industries color-code safety equipment locations, physical hazards, and protective equipment. Safety color codes were established by the American National Standards Institute (ANSI) and adopted by OSHA for use in hazardous areas. Porter Coatings' Safety Colors conform to OSHA and ANSI guidelines. OSHA does not specify the exact shade of color, but the color-coding should be consistent throughout a facility.*

☐ Red – The basic color for the identification of:
   o **Fire Protection Equipment and Apparatus.** Used for: fire alarm boxes, fire blanket boxes, fire buckets or pails, fire exit signs, fire extinguishers, fire hose locations, fire hydrants, fire pumps, fire sirens, post indicator valves for sprinkler system, and sprinkler piping.
   o **Danger.** Used for: safety cans or other portable containers of flammable liquids having a flashpoint at or below 80°F, table containers of flammable liquids (with additional clearly visible identification of the contents either in the form of a yellow band around the can or the name of the contents conspicuously stenciled or painted on the can in yellow), and danger signs.
   o **Stop.** Used for: emergency stop bars on hazardous machines and stop buttons or electrical switches used for emergency stopping of machinery.

☐ Orange – The basic color for designating dangerous parts of machines or energized equipment which may cut, crush, shock, or otherwise injure. Used to emphasize such hazards when enclosure doors are open or when gear belt or other guards around moving equipment are open or removed, exposing unguarded hazards.

☐ Yellow – The basic color for designating **caution**. Used for: marking physical hazards such as striking against, stumbling, falling, tripping, and "caught in between." Solid yellow, yellow and black stripes, yellow and black checkers (or yellow with suitable contrasting background) should be used interchangeably, using the combination that will attract the most attention in the particular environment.

☐ Green – The basic color for designating "safety" and the location of first aid equipment (other than fire-fighting equipment).

☐ Blue – The basic color for designating "caution," limited to warning against the starting, the use of, or the movement of equipment under repair or being worked on.

☐ Black, White, or combinations of Black and White. – The basic colors for designating traffic and housekeeping markings. Solid white, solid black, single color striping, alternate stripes of black and white, or black and white checkers should be used in accordance with local conditions.

☐ Identification of Contents of Piping Systems: *The American National Standard "Scheme for the Identification of Piping Systems" provides guidelines for the use of color on piping in plant facilities. Please consult 29 CFR 1910.144 and the ANSI publication A13.1-1975 for complete details.* Some general guidelines are:

- o Pipes are defined as conduits for the transport of gases, liquids, semi-liquids, or plastics.

- o This scheme does not cover pipes buried in the ground, or electrical conduits.

- o The Standard considers legend (written description of contents) to be primary. Color-coding is considered secondary.

- o Positive identification of the content of a piping system shall be by lettered legend giving the name of the contents in full or abbreviated form.

- o Arrows shall be used to indicate direction of flow.

- o Legends shall be applied close to valves and adjacent to changes in direction, branches, and where pipes pass through walls or floors, and at frequent intervals on straight pipe runs.

- o Identification may be accomplished by stenciling, the use of tape, or markers.

Color may be used to identify the characteristic properties of the contents as outlined in Table 1, using safety colors listed above.

Table 1

Materials Classification	Color of Field	Color of Letters for Legend
Hazardous Materials		
▶ Flammable, Explosive, Chemically active, Toxic, Extreme pressure or extreme temperature	Yellow	Black
▶ Radioactive	Purple	Yellow
Low Hazard Materials		
▶ Liquid or liquid mixture	Green	Black
▶ Gas or gaseous mixture	Blue	White
Fire Quenching Materials		
▶ Water, Foam, CO_2, etc.	Red	White

Where pipelines are located above the normal line of vision, the lettering shall be placed below the horizontal centerline of the pipe in such a manner as to make the legends visible to the eye.

HAZARD INDICES

☐ Chemical products throughout the employer's site should be labeled in accordance with the specifications of the written Hazard Communication Program. In some cases, manufacturer's labels are sufficient if the hazard of the product is appropriately communicated. The inclusion of numbers and letters to certain areas of these labels and/or the workplace chemical list describes the hazards and other information about the product. The tables below are intended as a reference guide to those codes. Employees and contractors should be trained to recognize these codes and protect themselves accordingly.

Table 2: Hazard Index Codes

Health Hazard (Blue)

Code	Type of Possible Injury
4	Life threatening, major or permanent damage may result from single or repeated exposures
3	Very dangerous; short exposure could cause serious, temporary, or residual injury
2	Temporary or minor injury may occur
1	Irritation or minor injury possible
0	No significant risk to health

Fire Hazard (Red)

Code	Type of Possible Injury
4	Extremely flammable, flash point below 0°F
3	Highly flammable, flash point 1°F-100°F
2	Moderately flammable, flash point 101°F-200°F
1	Slightly flammable, flash point above 201°F
0	Not flammable

Reactivity Hazard (Yellow)

Code	Type of Possible Injury
4	Readily capable of detonation at normal temperatures or pressures
3	Capable of detonation, but requires a strong initiating source, or reacts explosively with water
2	Normally unstable and may undergo violent chemical change, or react violently with water
1	Normally stable, but can become unstable at high temperatures or pressures, or may react with water
0	Stable material

Hazard Index - Basic Classifications

4	Severe Hazard
3	Serious Hazard
2	Moderate Hazard
1	Slight Hazard
0	Minimal Hazard

Table 3: Personal Protection Equipment (PPE) Codes for Working with Chemical Products

Code	Personal Protection Description
A	Safety glasses
B	Safety glasses, chemical gloves
C	Safety glasses, chemical gloves, synthetic apron
D	Face shield, chemical gloves, synthetic apron
E	Safety glasses, chemical gloves, dust respirator
F	Safety glasses, chemical gloves, synthetic apron, dust respirator
G	Safety glasses, chemical gloves, vapor respirator
H	Splash goggles, chemical gloves, synthetic apron, vapor respirator
I	Splash goggles, chemical gloves, dust and vapor respirator
J	Splash goggles, chemical gloves, synthetic apron, dust and vapor respirator
K	Airline hood or mask, chemical gloves, full suit, rubber boots
X	Ask your supervisor for special handling instructions

- [ ] All containers used in the storage and/or transportation of hazardous and toxic chemicals must be labeled, tagged, or marked with the following information:
  - o Identity of the hazardous chemical(s)
  - o Appropriate hazard warning
  - o The name and address of the manufacturer
  - o In some cases, this regulation is satisfied by the warning information that is printed on the product label. However, there are some products that do not have this information. In such cases, the employer/company will be responsible for identifying and labeling such products.

- [ ] Ensure that all portable containers are labeled. Portable containers are defined as pressure sprayers, spray bottles, squeeze bottles, hydraulic servicing units, or any container used in the transfer or temporary storage of products which contain hazardous ingredients as defined in Title 29 CFR 1910.1200.

- [ ] The Hazardous Materials Identification System (HMIS) is a standard system that communicates the hazards to the employee. The Hazardous Materials Identification Guide (HMIG), as it is referred to, communicates necessary hazard information by use of a common numerical system for Health (blue), Flammability (red), Reactivity (yellow), and Personal Protective Equipment required or specific hazard information (white).

> **NOTE:** If the product already contains the information compliant with this chapter, it is not necessary to repeat with an additional label.

- [ ] In order to properly complete the HMIS label (see Figure 2 for samples), it is necessary to reference the product information such as the MSDS or technical data. To save time, it is recommended to collect this data in a database or on/with the Workplace Chemical List. The Health, Flammability, and Reactivity is rated by using numerical ratings. To indicate the recommended Personal Protective Equipment, use the letter that is given in the PPE column. The definitions for these are found in Tables 2 and 3 above.

Figure 2: HMIG

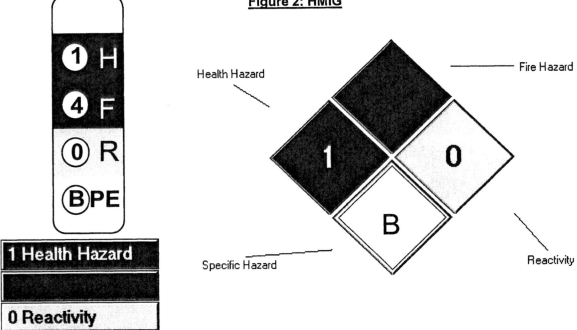

HEALTH HAZARD DESCRIPTION

☐ Materials that have one or more of the following characteristic properties are covered by the Hazard Communication Standard:

irritations	cutaneous hazards
highly toxic agents	corrosives
Blood-acting agents	toxic agents
hematopoietic system agents	eye hazards
sensitizers	carcinogens
hepatoxins	neurotoxins
nephrotoxins	reproductive toxins

agents that damage the lungs, skin, or mucous membranes

Health hazards are divided between acute health hazards and chronic health hazards. Some hazards have characteristics of both, but generally they are one or the other.

☐ Acute hazards – The effects of acute hazards are manifested soon after a single, brief exposure. Many acute effects disappear after a time and generally are not permanent. However, some may show permanent effects, and therefore can be considered both acute and chronic. *Acute, when used to describe a hazard, does not imply severity. For the purpose of this program, the term acute refers to a short-term effect.*

o ***Irritants*** - An irritant is defined as a chemical capable of causing reversible inflammation at the site of contact by chemical action. Examples include nitric oxide, sodium hypochlorite, stannic chloride, and ethyl alcohol.

o ***Cutaneous hazards*** - A cutaneous hazard is a material that affects the dermal layer of the body, such as; defatting of the skin, causing rashes or skin irritation. Examples include acetone and chlorinated compounds.

o ***Toxic agents*** - Toxic agents are those defined as substances which can cause acute injury to the human body, or which are suspected of being able to cause diseases or injury under certain conditions.

o ***Corrosive materials*** - A corrosive material causes visible destruction of or irreversible alterations in living tissue at the site of contact, by chemical reaction. Examples include caustic soda, sulfuric acid, hydrofluoric acid, phenol, and boron trifluoride.

o ***Eye hazards*** - An eye hazard is a material that affects the eye or visual capacity, for example, by causing conjunctivitis or corneal damage (organic solvents, acids, alkalis).

o ***Agents that act on the blood or hematopoietic system*** - This type of agent is a substance that decreases the hemoglobin function and deprives the body tissue of oxygen. Cyanosis and loss of consciousness are typical symptoms. Examples include carbon monoxide, cyanides, metal carbonyls, nitrobenzene, hydroquinone, and arsine.

☐ Chronic hazards – Chronic hazards have a long-term effect, essentially permanent. Their effects may be slow to develop, and often result from repeated or continuous exposure over a long period of time.

o ***Sensitizers*** - A sensitizer is a chemical that causes a substantial portion of exposed people or animals to develop an allergic reaction in normal tissue after repeated exposure. Examples include hydroquinone, bromine, platinum compounds, isocyanates, and ozone.

o ***Carcinogens*** - A substance or agent capable of causing or producing cancer in mammals, including humans. Examples include asbestos, benzene, beryllium, lead chromate, formaldehyde, vinyl chloride, trichloroethylene, carbon tetrachloride.

- o *Reproductive toxins (teratogens)* - Substances that can cause birth defects or sterility.

- o *Hepatoxins* - A hepatoxin is a chemical that can cause liver damage such as enlargement or jaundice. Examples include carbon tetrachloride, nitrosamines, vinyl chloride, chlorobenzene, trichloroethylene, chloroform, and ethyl alcohol.

- o *Nephrotoxins* - A nephrotoxin is a chemical that can cause kidney damage such as edema or proteinuria. Examples include halogenated hydrocarbons, uranium, vinyl chloride, trichloroethylene, and ethyl alcohol.

- o *Neurotoxins* - A neurotoxin is a chemical that causes primary toxic effects on the central nervous system, such as narcosis, behavioral changes, or decrease in motor functions. Examples include mercury, carbon disulfide, ethyl alcohol, acetylene, manganese, thallium, and tetraethyl lead.

- o *Agents that damage the lungs* - These agents irritate pulmonary tissue, resulting in cough, tightness in the chest, and shortness of breath. Examples include silica, asbestos, cotton fibers, coal dust, and toluene diisocyanate.

"RIGHT-TO-KNOW" POSTER

☐ A "Right-to-Know" poster must be displayed where products are regularly used that are considered hazardous by OSHA and state regulations. Posters should be displayed on safety bulletin board areas frequented by all affected employees and/or on affected shop bulletin boards.

RECORDKEEPING

☐ Training documentation.

☐ Retain (archive) all old or expired MSDSs.

---------- END OF SECTION ---------- ·

Section 18
Hazardous Chemicals
(Process Safety Management)

OVERVIEW

This section provides information regarding the requirements for preventing or minimizing the consequences of a release of toxic, reactive, flammable, or explosive chemicals. The releases may result in toxic, fire, or explosion hazards.

REGULATORY COMPLIANCE

☐ §1910.119 Process Safety Management of Highly Hazardous Chemicals

APPLICATION

☐ This section applies to processes with chemicals at or above a specific quantity. Refer to 29 CFR 1910.119 Appendix A for a list of chemicals and the threshold quantity for each chemical.

☐ This section applies to flammable liquids or gases on site in one location in quantities of 10,000 pounds or more **other than:**

 o Natural gas or fuel (gasoline or diesel) for comfort heating or vehicle refueling.

 o Flammable liquids in tanks that have controlled atmospheres and are stored at temperatures below their normal boiling point without chilling or refrigeration.

☐ This section does not apply to:

 o Retail facilities

 o Oil and gas drilling and/or servicing operations

 o Normally unoccupied remote facilities

DEFINITIONS

☐ ***Catastrophic release*** – major uncontrolled emission, fire, or explosion involving one or more highly hazardous chemical that could harm employees.

☐ ***Process*** – activities involving highly hazardous chemicals including use, storage, manufacturing, movement, or handling.

HAZARD ANALYSIS

☐ A hazard analysis is an orderly process used to obtain specific hazard, failure information, and data that pertain to the system or process that is being analyzed. This hazard analysis can be used to create a baseline for future monitoring and procedures or to identify problem areas.

☐ Employers shall retain process hazard analyses, updates, and re-evaluations for each process, for the life of the process.

☐ The employer shall complete a compilation of written process safety information before conducting any process hazard analysis required.

☐ The employer shall perform an initial process hazard analysis that identifies, evaluates, and controls the hazards involved in the process.

HAZARD ANALYSIS OF A PROCESS

☐ The employer shall determine the order of analyses based on the following considerations:
  o Extent of the potential hazard
  o Number of potentially affected employees
  o Age of the process
  o Accident history of the process

☐ Information contained in process hazard analysis is as follows:
  o Toxic information
  o Permissible exposure limits (PELs)
  o Physical data
  o Reactivity data
  o Corrosive data
  o Thermal and chemical stability data
  o Hazardous effects of accidental mixing of different materials that could occur

☐ Information required to conduct a process analysis includes:
  o Hazards of the process
  o Identification of any previous incident which had potential for severe consequences
  o Engineering and administrative controls to provide early warning of releases
  o Consequences of failure of engineering and administrative controls including the following:
    ◊ Facility setting
    ◊ Human factors
    ◊ Evaluation of the range of possible safety and health affects from failure of controls on employees in the work place

☐ Suggested methods to complete process hazard analysis include:
  o Block-flow diagram or process flow diagram

o   Process chemistry

o   Maximum intended inventory

o   Safe upper and lower limits for such items as temperature, pressure, flows, or compositions

o   An evaluation of the consequences of deviations to include those affecting the safety and health of employees

☐ Information pertaining to the equipment in the process shall include:

o   Materials of construction

o   Piping and instrumentation diagram

o   Electrical classification

o   Relief system design and design basis

o   Ventilation system design

o   Design codes and standards employed

o   Product material and energy balances for processes built after May 26, 1992

o   Safety systems such as interlocks, detection, or suppression systems

☐ Methods to determine and evaluate the hazards include:

o   Flow charts

o   What-if plans and checklists

o   Hazard and operability study (HAZOP)

o   Failure mode and effects analysis (FMEA)

o   Fault tree analysis

o   Any appropriate equivalent methodology

WRITTEN STANDARD OPERATING PROCEDURES (SOPs)

☐ The employer shall develop and implement written standard operating procedures (SOPs) that provide clear instructions for safely conducting activities involved in each covered process consistent with the process safety information.

☐ Operating procedures shall be made readily accessible to employees who work in or maintain a process.

☐ The operation procedures shall include the following:

o   Initial start-up date

o   Normal operating procedures

o   Normal shutdown

o   Temporary operation procedures

o   Emergency operation procedures

o   Start-up procedures for normal and emergency shutdowns

o   Emergency shutdown procedures which shall include:

◊   List of situations in which emergency shutdown is required

◊   Descriptions of the responsibilities of qualified operators

WRITTEN SAFE WORK PRACTICES

☐ The employer shall develop and implement safe work practices to provide for the control of hazards during operations such as these:

- o Lockout/Tagout

- o Confined space entry

- o Opening of equipment, pipes, or valves

- o Facility security control including:

    - ◊ Maintenance

    - ◊ Contractor

    - ◊ Laboratory

    - ◊ Support personnel

WRITTEN MAINTENANCE PROCEDURES FOR PROCESS EQUIPMENT

☐ Written preventive maintenance program for process equipment

☐ Inspection and testing of process equipment

☐ Inspection and testing of process equipment by manufacturers' recommendations or more frequently if determined by operating experience

☐ Documentation of each inspection and test on process equipment including the following information:

- o Name of inspector

- o Date of inspection and test

- o Serial number of equipment

- o Description of inspection and test

- o Results of inspection and test

TRAINING

☐ Each employee presently involved in operating a process, and each employee before being involved in operating a newly assigned process, shall be trained in the overview of the process and in the operating procedures.

☐ Initial training shall include:

- o Emphasis on the specific safety and health hazards

- o Emergency operations including:

    - ◊ Shutdown

    - ◊ Safe work practices

    - ◊ Incident reporting

☐ Instead of the initial training for those employees already operating a process as of May 26, 1992, an employer may certify in writing that the employee has the required knowledge, skills, and abilities to safely carry out the duties and responsibilities as specified in the operating procedures.

- [ ] Refresher training shall be provided at least every 3 years and more often if necessary. The employer, in consultation with the employees involved in the operating process, shall determine the appropriate frequency of refresher training.

- [ ] The employer shall retain a record that contains the identity of the employee, date of training, and means used to verify that the employee understood the training.

RECORDKEEPING

- [ ] Inspections and tests (daily, monthly, annual)

- [ ] Maintenance records and procedures

- [ ] Training

- [ ] Written hazard analysis plan

- [ ] Written standard operating procedures (SOPs)

- [ ] Process safety information

- [ ] Analysis

- [ ] Monitoring methods

- [ ] Process hazard analysis

- [ ] Process equipment analysis

- [ ] Process safety procedures

- [ ] Process safety practices

---------- END OF SECTION ----------

Section 19
Hazardous Waste Operations and Emergency Response (HAZWOPER)

OVERVIEW

This section provides the requirements for employers to follow during hazardous waste cleanup, containment, and emergency response.

REGULATORY COMPLIANCE

☐ §1910.120 Hazardous Waste Operations and Emergency Response

SAFETY AND HEALTH PROGRAM

☐ Employers must develop a written safety and health program for employees involved in hazardous waste operations.

☐ The safety and health program must include:

   o   Organizational Structure

   o   Comprehensive Work Plan

   o   Site-Specific Safety and Health Plan

   o   Safety and Health Training Program

   o   Medical Surveillance Program

   o   Standard Operating Procedures (SOPs)

   o   Necessary Interface between General Program and Site-Specific Activities

SITE CHARACTERIZATION AND ANALYSIS

☐ Hazardous waste sites shall be evaluated for site hazards and to determine the appropriate safety and health control procedures needed to protect employees from identified hazards.

☐ A qualified person shall perform a preliminary evaluation of the site's characteristics in order to determine protection methods prior to entry.

☐ All suspected hazards that may pose inhalation or skin absorption hazards shall be identified during the survey.

- [ ] Required information:
  - o Location and approximate size of the site
  - o Description of the response activity and/or the job task to be performed
  - o Duration of planned employee activity
  - o Site topography and accessibility by air and roads
  - o Safety and health hazards expected at the site
  - o Pathways for hazardous substance dispersion
  - o Status of emergency response teams
  - o Chemical and physical hazards expected

- [ ] Personal protective equipment required at the site.

- [ ] Employee exposure monitoring

SITE CONTROL

- [ ] A site control program shall be developed during the cleanup planning activities.

- [ ] Elements of the site control program include:
  - o Site map
  - o Site work zones
  - o Use of "buddy system"
  - o Site communications
  - o Nearest medical assistance
  - o SOPs and safe work practices

TRAINING

- [ ] All personnel on the site shall be trained in hazardous waste operations before participating in any activity that could expose them to hazardous substances, or safety or health hazards.

- [ ] Training should consist of the following elements:
  - o Names of persons responsible for site safety and health operations
  - o Safety, health, and other hazards present on the site
  - o Use of personal protective equipment
  - o Safe work practices
  - o Safe use of engineering controls and equipment
  - o Medical surveillance requirements, including recognition of symptoms and signs of overexposure to hazards
  - o Decontamination procedures
  - o Emergency response procedures
  - o Confined-space entry procedures
  - o Spill containment procedures

☐ General site workers, laborers, and supervisors shall have a minimum of 40 hours offsite instruction and 3 days onsite training under the direct supervision of a trained, experienced supervisor (HAZWOPER).

☐ Workers on the site occasionally and workers regularly on site shall receive at least 24 hours of offsite instruction and 1 day onsite training by a trained, experienced supervisor.

☐ Regular workers required to wear respirators shall undergo an additional 16 hours of offsite instruction and 2 days onsite training by a trained, experienced supervisor.

☐ Management personnel and supervisors shall attend at least 40 hours offsite instruction and 3 days of field-supervised training and an additional 8 hours of specialized training on topics such as personal protective equipment, employee training, spill containment, and monitoring techniques.

☐ Trainers shall be qualified to instruct employees and must have completed a trainer's course and attained certification as a trainer from that course, or they should have academic credentials and instructional experience necessary for teach the subjects.

☐ Each certified worker shall undergo an additional 8-hour refresher course annually.

MEDICAL SURVEILLANCE

☐ A medical surveillance program must be implemented for employees participating in this program who work with hazardous substances.

ENGINEERING CONTROLS AND WORK PRACTICES

☐ When a hazard is encountered, employers must utilize engineering controls as the primary method to protect employees on the job.

☐ If engineering controls are not feasible, employers will implement work practices or administrative controls to limit employee exposure on the job.

☐ Personal protective equipment (PPE) is the last resort once attempts to implement engineering controls and work practices have been exhausted.

MONITORING

☐ Air monitoring shall be performed initially and periodically when employees are potentially exposed to hazards at or above IDLH (Immediately Dangerous to Life and Health) atmospheres.

OTHER ISSUES TO CONSIDER

☐ Informational programs for employees.

☐ Handling drums and containers.

☐ Decontamination procedures.

☐ Emergency response by employees at uncontrolled waste sites.

☐ Illumination plans (lighting for worksites).

☐ Sanitation at temporary workplaces.

☐ New technology programs (training methods, foams, chemical neutralizers, etc.).

☐ Operations conducted as part of the Resource Conservation and Recovery Act (RCRA) of 1975.

☐ Emergency response program.

---------- END OF SECTION ----------

Section 20
Ladders

OVERVIEW

This section provides guidelines for the use, maintenance, and inspection of industrial ladders. This section also identifies the marking requirements for different types of ladders used in the workplace today.

REGULATORY COMPLIANCE

☐ §1910.25 Portable Wooden Ladders

☐ §1910.25 Portable Metal Ladders

☐ §1910.25 Fixed Ladders

GENERAL REQUIREMENTS

☐ Ensure steps have a minimum load capacity of 250 pounds.

☐ Inspect ladders for damage prior to use.

☐ Do not place ladders against movable or unstable objects.

☐ When possible, secure ladders at the top and bottom to prevent movement while in use.

☐ Clean ladder feet of mud, grease, or other substances that could cause a slip or fall.

☐ Do not place ladders on unstable bases such as boxes or barrels.

☐ Do not stand on the top two steps of a stepladder.

☐ Ladders must extend at least 3 feet above the point of support, at eaves, gutters, rooflines, etc.

☐ Fully open stepladders to permit the spreaders to lock.

☐ Ensure all labels are in place and legible.

☐ Move the ladder as required to avoid overreaching.

☐ Single ladders must not be more than 30 feet in length.

☐ Extension ladders up to 36 feet must have a 3-foot overlap between sections.

☐ Extension ladders over 36 feet and up to 48 feet shall have a 4-foot overlap between sections.

☐ Extension ladders over 48 feet and up to 60 feet shall have a 5-foot overlap between sections.

☐ Two-section extension ladders shall not exceed 48 feet in total length. Ladders with more than two sections shall not exceed 60 feet in total length.

☐ Do not use ladders horizontally as scaffolds, runways, or platforms.

☐ Keep the areas at the top and base of ladders free of tripping hazards such as loose materials, trash, cords, hoses, welding leads, etc.

☐ Place the base of straight or extension ladders approximately 1/4 of the working length of the ladder from the vertical axis.

☐ Do not allow ladders to project into passageways or doorways where they could be struck by personnel, moving equipment, or materials being handled without protection by barricades or guards.

☐ Employees must face the ladder when ascending or descending.

☐ Employees must use both hands when going up or down a ladder. Use ropes to lift loads.

☐ Provide ladder safety training to all employees.

MAINTENANCE OF LADDERS

☐ Authorized manufacturer representatives shall perform repairs.

☐ Conduct inspections before each use.

☐ Defective, broken, or damaged ladders must be removed from service and tagged accordingly.

☐ The rungs must be tight in the joints of the side rails.

☐ All moving parts must operate freely without binding.

☐ Lubricate all pulleys, wheels, and bearings frequently.

☐ Keep rungs free of grease and oil.

☐ Replace immediately ropes that are badly worn, frayed, or damaged.

☐ All ladders must be equipped with slip-resistant feet, free of grease, and in good condition.

PORTABLE WOODEN LADDERS

☐ Keep all wood ladders free of splinters, sharp edges, shake, wane, compression failures, decay, and other irregularities.

☐ Portable stepladders must be no longer than 20 feet.

☐ The steps must be spaced no more than 12 inches apart.

☐ Stepladders must have a metal spreader or locking device of sufficient strength and size to steady the front and back legs of the ladder when open.

PORTABLE METAL LADDERS

☐ Inspect ladders immediately before use, when dropped, or when tipped over.

☐ Metal ladders must not be used for electrical work or in areas where they could contact energized wiring.

FIXED LADDERS

☐ Steps must be no more than 12 inches apart.

☐ Site-made ladders must be constructed to conform to the established OSHA standards.

☐ All fixed ladders must be painted or treated to prevent rusting.

☐ Fixed ladders 20 feet or higher must have a landing every 20 feet if there is no surrounding cage. If the ladder has a cage or safety device, a landing is required every 30 feet.

---------- END OF SECTION ----------

Section 21
Lockout/Tagout Program

OVERVIEW

This section describes the requirements for locking out equipment that could become energized or contain stored energy that might injure employees during cleaning or maintenance. OSHA requires the development of a written Lockout/Tagout (LOTO) program (herein referred to as Lockout Program) to prevent the unexpected release of energy that could cause injury. This section applies to all areas where employees may be exposed to or work on equipment that could become energized or contain stored energy that might injure employees.

The employer must determine if his/her employees are subject to this program. Some general guidelines for this determination are:

- o   Those individuals who work in building and equipment maintenance.

- o   Employees who repair shop equipment.

- o   Employees required to service equipment or clear jams by entering areas with potentially rotating, cutting, or pounding assemblies or parts.

- o   Facilities or maintenance department employees who work on or around affected machines or equipment.

This standard does not apply to work on cord-and plug-connected electric equipment for which exposure to the hazards of unexpected energization or start up of the equipment is controlled by the unplugging of the equipment from the energy source and by the plug being under the exclusive control of the employee performing the servicing or maintenance.

REGULATORY COMPLIANCE

☐   §1910.147 The Control of Hazardous Energy (Lockout/Tagout)

☐   American National Standards Institute (ANSI) Z244.1-1982

☐   Local government regulations

TERMS AND DEFINITIONS

☐ **_Authorized Agent_** – An authorized agent is a person who has been trained to implement energy controls (locks) to ensure their safety while servicing or performing maintenance on that machine or equipment. For areas that utilize _Authorized Agents,_ a list of these qualified individuals must be posted so _Affected Agents_ know whom to contact when needed.

☐ **_Affected Agent_** – An employee whose job requires him/her to operate or use a machine or equipment on which servicing or maintenance is being performed under lockout, or whose job requires him/her to work in an area in which such servicing or maintenance is being performed. Affected agents must be able to recognize and understand that locks are designed to prevent any attempt to restart or re-energize equipment.

☐ **_Energy Source_** – An energy source may be electrical, hydraulic, thermal, or pneumatic. Also included is the energy stored in springs and electrical capacitors and the potential energy from suspended parts.

☐ **_Lockout (LO)_** – A term that refers to a system of procedures and training used to notify and remind employees to trace and disengage all energy sources that could cause machinery/equipment to cycle or start during servicing and/or maintenance. An energy source may be electrical, hydraulic, thermal, or pneumatic. Also included is the energy stored in springs and electrical capacitors and the potential energy from suspended parts.

☐ **_Lockout Device_** – A device that utilizes a positive means such as a lock to hold an energy-isolating device in the safe position and prevents the energizing of a machine or equipment. Locks must be standardized and easily identifiable as lockout devices either by color, shape, or size. Locks must be substantial in order to prevent removal, except by excessive force from special tools such as bolt cutters or other metal cutting tools. Locks must clearly identify the employee who applies them.

☐ **_Supervisor_** – For the purposes of this section, supervisor is interchangeable with the terms director, manager, lead, and/or safety representative.

☐ **_Tag(s)_** – A prominent warning device and a means of attachment, which is securely fastened to an energy isolating device (lockout device, or lock) in accordance with an established procedure, to indicate the name of the person applying the tag and that the energy isolating device (LO device) shall not be removed and the equipment being controlled may not be operated. Tags may only be removed by the person identified on the tag or by his/her supervisor acting under a tag removal procedure. Tags shall never be bypassed, ignored, or otherwise defeated. Tags shall be used with each energy-securing lock/device installed.

Examples:
DO NOT START   DO NOT OPEN
DO NOT CLOSE   DO NOT OPERATE

TRAINING AND QUALIFICATIONS

☐ Training Programs - The employee training program shall include training in the recognition of applicable hazardous energy sources, the type and magnitude of the energy available in the workplace, and the methods and means necessary for energy isolation and control.

   o Authorized Employee: The employer shall initiate *authorized* employee LO training programs to cover specific procedural steps for shutting down, isolating, blocking, and controlling the types of hazardous energy with which his/her shop or area is likely to come into contact when performing maintenance on equipment or clearing conveyor jams. Newly hired and transferred employees shall be trained on arrival to an affected shop or area. Recurrent training shall be conducted under any one of the following conditions:

      ◊ A new energy hazard is introduced in the individual's work operation

      ◊ New equipment is added to the area

      ◊ The area's LO procedures are changed

      ◊ The individual demonstrates the need based on action or knowledge.

   o Affected Employee: The employer shall initiate *affected* employee LO training for all employees who may come in contact with or operate equipment that may be locked out. To certify that the individual has received training, the trainee's name and date of training shall be kept on file locally. Documentation for those trained and authorized for LO shall be maintained (hard copy or electronically) and readily available for inspection by government officials or company auditors.

MINIMUM REQUIREMENTS FOR LOCKOUT PROCEDURES

☐ This procedure establishes the minimum requirements for LO of energy sources that could cause injury to personnel. Written energy control procedures shall be completed for each piece of equipment that requires LO, unless the piece of equipment meets the documentation exception test described in the "Lockout Procedures" Subsection (third bullet).

☐ Rules for LO – All equipment shall be locked out to protect against accidental or inadvertent operation when such operation could cause injury to employees. Do not attempt to operate any switch, valve, or other energy-isolating device bearing a lock. Areas having a need to operate locked out equipment shall observe the instructions on the tag and coordinate with the Authorized Agent.

☐ Preparation for LO – Employees *authorized* to perform LO shall be certain as to which switch, valve, or other energy-isolating devices applies to the equipment being locked out. More than one energy source (electrical, mechanical, etc.) may be involved. Employees shall clear any questionable identification of sources with their supervisor.

☐ Sequence of Lockout Procedures – Notify all *affected* employees and departmental management that a LO is required and the reason required.

   o If the equipment is operating, shut it down by the normal stopping procedure (depress stop button, open toggle switch, etc.).

   o Engage the switch, valve, or other energy-isolating device so the energy sources (electrical, mechanical, hydraulic, etc.) are disconnected or isolated from the equipment. Stored energy, such as that found in capacitors, springs, elevated machine members, rotating flywheels, hydraulic systems, and air, gas, steam, or water pressure, etc., must also be dissipated or restrained by methods such as grounding, repositioning, blocking, bleeding down, etc.

   o Lock out the energy-isolating devices with an assigned individual lock.

o   After ensuring that no employees are exposed to the machinery and/or equipment and to ensure that the energy sources are properly disconnected, operate the push-button or other normal operating controls to make certain the equipment will not operate.

CAUTION: Return operating controls to the neutral (off) position after the test.

o   The equipment is now locked out.

☐ Restoring Equipment to Service – The following procedures may be used when the job is complete and the equipment ready for testing or normal service only by the person who applied the lock.

o   Check the equipment to see that the equipment is clear of employees.

o   Verify all actuating devices are in the neutral (off) position.

o   Remove all locks.

o   Remove energy-isolating devices.

o   After the lockout device is removed and before equipment is started, affected employees shall be notified that the LO device has been removed and the equipment is about to be started.

o   Operate to verify energy has been restored to equipment.

☐ Procedures Involving More than One Person – If more than one individual is required to perform services requiring LO on equipment, each shall place his own personal lock on the energy-isolating devices. A multi-lock hasp must be used when more than one lock is applied.

☐ Contractors – Organizations interfacing with outside contractors performing service or maintenance requiring LO shall ensure that the contractors provide information relative to their LO program. Affected employees shall be trained on the types of locks and tags used by the contractor and to not remove them or attempt to operate the equipment. The employer shall also inform contractors of the company's LO procedures.

☐ Cross-functional Work – Organizations providing service requiring LO shall ensure the employees affected by their service are knowledgeable about the LO process being used.

☐ Lock Removal – In the event an authorized employee is not available to remove his/her lock, a supervisor may remove it after the following actions have been taken:

o   A check has been made of the duty roster to verify the employee is no longer on site.

o   Attempts to contact the employee by telephone, radio, or pager shall be made to inform the employee that his/her lock is being removed.

o   If verbal contact cannot be made with the employee, a sign/note that his/her lock has been removed will be posted on his/her locker/toolbox/desk/workbench or other suitable location that the employee will check upon returning to the work area.

o   Only after the above steps have been taken and documented can a LO lock be removed by either using the supervisor controlled extra key or by cutting off the lock.

☐ Change of Shift – The lock transfer process shall ensure that each employee removes his/her lock and tag prior to leaving the station for the day (only after the next person locks and tags out). If locks must be left in place over shift change, the new shift must be informed of locks and reason for continued LO condition.

☐ Security Lock – A security lock that more than one person has a key to may be used to secure equipment from unintended operation while it is unattended but not as a LO lock.

☐ <u>Extra Keys</u> – The employer may include a procedure to allow for a department head (Facilities or Maintenance Manager) or supervisor to control extra keys only after the steps in the "Lock Removal" Subsection above have been satisfied. **BP**: This practice is not recommended.

PERIODIC OBSERVATIONS

☐ At least annually, an authorized employee, other than the one utilizing equipment's energy control procedure (operators cannot observe themselves) is required to observe and verify the effectiveness of equipment energy control procedures. These observations shall at least provide for a demonstration of the procedures and may be implemented through random audits and/or planned visual observations. These observations are intended to ensure the energy control procedures are being properly implemented and to provide an essential check on the continued utilization of the procedures.

☐ The observations shall include a review with all authorized employees who use equipment or machinery, of their responsibilities under the energy control procedure being inspected. Group meetings between the authorized employee who is performing the observation and all authorized employees who implement the procedure would constitute compliance with this requirement. Periodic observations must provide for and ensure effective correction of identified deficiencies. Finally, the employer is required to certify that the prescribed periodic observations have been performed through a record that is kept on file locally for 2 years. These records shall identify all of the following:

o The machine or equipment on which the energy-control procedures were being utilized

o The date of the observation

o The individual(s) included in the observations

o The individual(s) performing the observations

RESPONSIBILITY AND ACTIONS

BP: This subsection is not part of the Standard, but included here to help identify areas within facilities that must be included in a Lockout Program.

Responsibility of all Operation Departments (Production, Machine Shop, etc.):

<u>Action 1</u>: Survey all jobs in your group and determine if company employees do service or perform maintenance work as part of their job. Document survey results. If survey results show no LO program required go to Action 3.

<u>Action 2</u>: If survey results show employees do service or maintain equipment, supervisors must ensure those employees are trained as Authorized Agents.

<u>Action 3</u>: For all employees who do not perform maintenance on, but use equipment affected by the LO program, ensure employees complete training for *affected* employees, where awareness of the LO program is taught. Ensure that other agencies performing service and maintenance requiring LO in your area have provided information on their LO programs for your employees. Ensure that the company's LO program is shared with contract services and other non-company employees working in the area.

<u>Action 4</u>: Prohibit all employees not trained as *authorized* agents from performing service or maintenance on equipment that falls under the LO program.

LOCKOUT PROCEDURES

☐ <u>Preparation for LO</u> – Employees authorized to perform LO shall be certain as to which switches, valves, or other energy-isolating devices apply to the equipment being locked out. More than one energy source (electrical, mechanical, etc.) may be involved. Only with help from a supervisor shall employees clear unknown or questionable sources of energy.

☐ <u>Identifying the LO Procedure</u> – Procedures must be completed for each piece of affected equipment or machinery. Contact the program administrator for additions of or modifications to equipment that will need LO procedures developed.

BP: The program administrator will develop the lockout procedures and a placard (if required by the LO program) for the equipment. The placard should include:

- o The equipment ID number, type of equipment, serial number, and location
- o Maintaining department's contact number
- o The specific types of energy controlled, identified by magnitude and type of energy
- o Lockout location and method
- o Release of stored energy procedures
- o Verification of isolation
- o Any special notes

☐ <u>Exception Test</u> – Documentation of LO procedures will not be required if **all** of the following elements exist:

- o The equipment has no potential for stored or residual energy, and there is no chance of energy re-accumulation after shut down which could endanger employees.
- o The equipment has a single energy source that can be readily identified and isolated.
- o The isolation and locking out of the equipment's energy source will completely de-energize and deactivate the equipment.
- o The equipment is isolated from its energy source and locked out during servicing or maintenance.
- o A single control source will achieve a locked-out condition.
- o The control source will be under the exclusive control of the *authorized* employee performing the service or maintenance on the equipment.
- o Servicing does not create hazards for other employees.
- o There have been no accidents involving the unexpected activation or re-energizing of this equipment during servicing or maintenance.

RECORDKEEPING

☐ Recordkeeping

- o Maintain a record of authorized employees
- o Training documentation
- o Periodic inspections/observations
- o Written LO procedures (document for each applicable piece of equipment)

---------- END OF SECTION ----------

Section 22
Machinery & Machine Guarding

OVERVIEW

The purpose of this section is to provide general requirements for machine guarding standards, procedures, engineering controls, and inspections. Machine safety practices are an integral part of any safety program, and suggestions to make any machine safer should always be welcomed.

REGULATORY COMPLIANCE

☐ §1910.212 General Requirements for All Machines

☐ §1910.213 Woodworking Machinery Requirements

☐ §1910.215 Abrasive Wheel Machinery

☐ §1910.216 Mills And Calendars in the Rubber and Plastics Industries

☐ §1910.217 Mechanical Power Presses

☐ §1910.218 Forging Machines

☐ §1910.219 Mechanical Power-Transmission Apparatus

TRAINING QUALIFICATIONS

☐ Before starting work on any operation, the employer shall ensure the operator is thoroughly familiar with the safe methods of work using machinery for the operation. The employer can accomplish this through adequate supervision ensuring that correct operating procedures are being followed.

MACHINE GUARDING GENERAL

☐ Types of Guarding – One or more methods of machine guarding shall be provided to protect operators and other personnel in the machine area from hazards such as those created by:

- o Point of operation
- o In-going nip points
- o Rotating parts
- o Flying chips
- o Sparks

Examples of guarding methods include barrier guards, two-hand tripping devices, and electronic safety devices.

☐ <u>Point of Operation</u>

o   The point of operation is the area on a machine where work is actually performed upon the material being processed. Machines that expose a person to injury shall be guarded. The guarding shall be designed and constructed to prevent the operator from having any part of his body in the danger zone during the operating cycle.

o   Special hand tools for placing/removing materials shall permit easy handling of material without the operator placing a hand in the danger zone. Such tools shall not be in lieu of required guarding, but only to supplement protection.

o   Following are some of the machines that usually require point of operation guarding:

◊   Guillotine cutters

◊   Shears

◊   Alligator shears

◊   Power presses

◊   Milling machines

◊   Power saws

◊   Jointers

◊   Portable power tools

◊   Forming rolls and calendars

☐ Revolving drums, containers, and barrels shall be guarded by an enclosure interlocked with the drive mechanism so the drum, container, or barrel cannot revolve unless the guard enclosure is in place.

☐ When the periphery of the blades of a fan is less than 7 feet above the floor or working level, the blades shall be guarded. The guard shall have no openings larger than 1/2 inch.

☐ Machines designed for a fixed location shall be securely anchored to prevent "walking" or moving. Machines that are so heavy that they will not "walk" are exempted from this requirement.

☐ <u>Machine Controls and Equipment</u>

o   Machines shall have mechanical or electrical power controls to allow the operator to cut off power at the point of operation without leaving his position at the point of operation.

o   On machines driven by belts and shafting, a locking type belt shifter or an equivalent positive device shall be used.

o   Machines shall be prevented from automatically restarting after power failures if there is a possibility the operator might be injured.

o   Power controls and operating controls should be within easy reach of the operator at the regular work location. It should not be necessary for the operator to reach over the cutter to make adjustments.

WOOD-WORKING MACHINERY REQUIREMENTS

☐ <u>Hand-Fed Ripsaws</u> – A hood completely enclosing the portion of the saw above the table and the portion of the material being cut shall guard circular hand-fed ripsaws.

- o The hood and mounting shall be arranged so the hood will automatically adjust itself to the thickness of and remain in contact with the material being cut.

- o The hood will not offer any considerable resistance to the insertion of material to the saw or to the passage of the material being sawed.

- o The hood shall be strong enough to resist blows and strains of reasonable operation, adjusting, and handling.

- o The hood shall protect the operator from flying splinters and broken saw teeth.

- o The hood shall be made of a material soft enough to avoid causing saw tooth breakage.

- o The hood shall be mounted to ensure positive, reliable operation in true alignment with the saw.

- o The mounting shall be strong enough to resist any reasonable side thrust or other force tending to throw it out of line.

- o Hand-fed circular ripsaws shall have a spreader to prevent material from squeezing the saw or being thrown back on the operator. A spreader is not required for grooving, dadoing, or rabbeting. The spreader shall be immediately replaced upon completion of the operation.

- o Hand-fed circular ripsaws shall have non-kickback fingers or dogs to oppose the thrust or tendency of the saw to pick up the material or to throw it back toward the operator. The non-kickback fingers shall provide adequate holding power for all material thicknesses being cut.

☐ <u>Hand-Fed Crosscut Table Saws</u> – Circular crosscut table saws shall be guarded by a hood and have a spreader to meet the above requirements.

☐ <u>Circular Saws Shall</u>:

- o Be guarded by a hood or shield of metal above the saw to guard against flying splinters or broken saw teeth.

- o Have a spreader fastened securely behind the saw. The spreader shall be slightly thinner than the saw kerf and slightly thicker than the saw disk.

  Exception: Self-feed saws with rollers or wheels at the back of the saw.

☐ <u>Swing/Sliding Cutoff Saws</u>

- o Swing and/or sliding cutoff saws shall have a hood completely enclosing the upper half of the saw, the arbor end, and the point of operation at all positions of the saw.

- o Swing and/or sliding cutoff saws shall be able to return the saw automatically to the back of the table when released at any point.

- o Limit chains or other equally effective devices shall prevent the saw from swinging or sliding beyond the front or back edges of the table.

- o Inverted swing cutoff saws shall have a hood to cover the part of the saw that protrudes above the top of the table or above the material being cut. The hood shall automatically adjust itself to the thickness of and remain in contact with the material being cut.

☐ Radial Saws

- o The upper hood shall completely enclose the upper portion of the blade down to include the saw arbor. The sides of the lower exposed portion of the blade shall be guarded to the full diameter of the blade by a device that will automatically adjust itself to the thickness of the material and remain in contact with the material being cut to give maximum protection possible for the operation being performed.

- o Radial saws used for ripping shall have non-kickback fingers or dogs on both sides of the saw to oppose the thrust or tendency of the saw to pick up the material or to throw it back toward the operator. The non-kickback fingers or dogs shall provide adequate holding power for all material thicknesses that will be cut.

- o An adjustable stop shall prevent the forward travel of the blade beyond the position necessary to complete the cut in repetitive operations.

- o The saw shall be installed so the front end of the unit is slightly higher than the rear, to cause the cutting head to return gently to the starting position when released by the operator.

- o Ripping/cutting shall be against the direction in which the saw rotates. The direction of rotation shall be conspicuously marked on the hood. A permanent label colored standard danger red, at least 1-1/2 inches by 3/4 inch shall be affixed to the rear of the guard reading "Danger, Do Not Rip/Cut From This End."

☐ Band Saws

- o All portions of the saw blade shall be enclosed or guarded, except for the working portion of the blade between the bottom of the guide rolls and the table.

  - ◊ Band saw wheels shall be fully encased.

  - ◊ The outside periphery of the enclosure shall be solid.

  - ◊ The front and back of the band wheels shall be enclosed either by solid material, wire mesh, or perforated metal.

    - ▪ Mesh or perforated metal shall not be less than 0.037 inch (U.S. Gage #20) and the openings shall not be greater than 3/8 inch.

    - ▪ Solid materials shall be of an equivalent strength and firmness.

  - ◊ The guard for the portion of the blade between the sliding guide and the upper saw-wheel guard shall protect the saw blade at the front and outer side. This portion of the guard shall be self-adjusting to raise and lower with the guide.

  - ◊ The upper wheel guard shall be made to conform to the travel of the saw on the wheel.

- o Band saw machines shall have a tension control device to indicate a proper tension for the standard saws used on the machine to assist in eliminating saw breakage due to improper tension.

☐ Jointers

- o Hand-fed planers and jointers with horizontal heads shall have a cylindrical cutting head, with a knife projection not exceeding 1/8 inch or less beyond the cylindrical body of the head.

- o The opening in the table shall be kept as small as possible. The clearance between the rear table edge and the cutter head shall be 1/8 inch or less. The table throat opening shall be 2-1/2 inches or less when the tables are set or aligned with each other for zero cut.

- o Hand-fed jointers with horizontal cutting heads shall have an automatic guard covering all of the section of the head on the working side of the fence or cage. The guard shall automatically adjust itself to cover the unused portion of the head and shall remain in contact with the material at all times.

- o Hand-fed jointers with horizontal cutting heads shall have a guard covering the section of the head in back of the cage or fence.

- o Wood jointers with vertical heads shall have either an exhaust hood or other guard enclosing the revolving head completely, except for a slot wide enough for the application of the material to be jointed.

☐ Maintaining Woodworking Machinery:

- o Dull, badly set, improperly filed, or improperly tensioned saws shall be removed from service immediately. Saws to which gum has adhered on the sides shall be cleaned immediately.

- o All knives and cutting heads of woodworking machines shall be kept sharp, properly adjusted, and firmly secured. Two or more knives used in one head shall be properly balanced.

- o Bearings shall be well lubricated and kept free from lost motion.

- o Arbors of all circular saws shall be free from play.

- o Only skilled individuals shall do sharpening or tensioning of saw blades or cutters.

- o All cracked saws shall be removed from service.

- o Inserting wedges between the saw disc and the collar to form a "wobble saw" shall not be permitted.

- o Push sticks or push blocks suitable for the job shall be provided.

- o The knife blade of jointers shall be installed and adjusted to protrude not more than 1/8 inch beyond the cylindrical body of the head.

- o The area around woodworking machinery shall be kept clean to ensure that guards function effectively to prevent fire hazards in switch enclosures, bearings, and motors.

ABRASIVE WHEEL MACHINERY

☐ Abrasive wheels shall be used only on machines provided with safety guards.

Exceptions:

- o Wheels used for internal work (not exposed) within the component being ground.

- o Mounted wheels used in portable operations that are 2 inches or smaller in diameter.

☐ Guard Design

- o The safety guard shall cover the spindle end, nut, and flange projections and shall be mounted to maintain proper alignment with the wheel.

- o The angular exposure of the grinding wheel periphery and sides for safety guards used on bench and floor stands should not exceed 90° or 1/4 of the periphery.

- o This exposure shall begin not more than 65° above the horizontal plane of the wheel spindle.

- o The exposure shall not exceed 125° when in contact with the wheel below the horizontal plane of the spindle.

- [ ] <u>Work Rests</u> – Work rests shall be used on bench grinding machines to support the work.
  - o They shall be of rigid construction and adjustable to compensate for wheel wear.
  - o They shall have a maximum opening of 1/8 inch to prevent wheel breakage caused by the work being jammed between the wheel and the rest.
  - o The work rest shall be securely clamped after each adjustment.
  - o Adjustments shall not be made while the wheel is in motion.

- [ ] <u>Peripheral Guards</u>
  - o Peripheral guards shall be constructed so the peripheral protecting member can be adjusted to the constantly decreasing diameter of the wheel.
  - o The distance between the guard and the wheel shall never exceed 1/4 inch.
  - o Adjustments shall not be made while the wheel is in motion.

- [ ] <u>Flanges: General Requirements</u> – All abrasive wheels shall be mounted between flanges not less than 1/3 the diameter of the wheel.

  Exceptions:
  - o Mounted wheels
  - o Portable wheels with threaded inserts or projecting studs
  - o Abrasive discs (inserted nut, inserted washer, and projecting stud type)
  - o Plate-mounted wheels
  - o Cylinders, cups, or segmental wheels mounted in chucks

- [ ] Flanges shall be dimensionally accurate and in good balance, without rough surfaces or sharp edges.

- [ ] Flanges of any type between which a wheel is mounted shall be of the same diameter and have equal bearing surface. Certain wheels, because of their shape and usage, require flanges with specially designed adapters.

- [ ] Blotters (compressible washers) shall always be used between flanges and abrasive wheels to ensure uniform distribution of flange pressure.

  EXCEPTIONS:
  - o Mounted wheels
  - o Abrasive discs
  - o Plate-mounted wheels
  - o Cylinders, cups, or segmental wheels mounted in chucks
  - o Certain type of cutting-off wheels and internal wheels

- [ ] <u>Repairs and Maintenance</u> – All flanges shall be maintained in good condition. When the bearing surfaces become worn, warped, sprung, or damaged, they shall be trued, refaced, or replaced.

☐ <u>Mounting: Inspection, Sounding</u> – The user shall closely inspect and sound by ring test all wheels immediately before mounting to make sure they have not been damaged in transit, storage, or otherwise. Wheels should be tapped gently with a nonmetallic implement such as a screwdriver handle for light wheels or a wooden mallet for heavier wheels.

- o Wheels are tapped about 45° on each side of the vertical centerline and about 1 or 2 inches from the periphery.

- o The wheels should be rotated to repeat the test.

- o Wheels must be dry and free from sawdust before applying the test

> **NOTE:** A sound, undamaged wheel gives a clear metallic tone. A cracked wheel will have a "dead" sound.

- o The spindle speed of the machine shall be checked before mounting the wheel to be certain it does not exceed the maximum operating speed marked on the wheel.

☐ <u>Arbor Size</u> – Grinding wheels shall fit freely on the spindle and remain free under all grinding operations.

☐ All contact surfaces of wheels, blotters, and flanges shall be flat and free of foreign matter.

☐ A bushing shall not exceed the width of the wheel and shall not come in contact with the flanges.

☐ After mounting a wheel, care should be taken to see that the safety guard is properly positioned (per manufacturer's specifications) before starting the wheel.

MECHANICAL POWER PRESSES

☐ These requirements shall apply to the maintenance and use of all mechanical power presses, with these exceptions:

- o Press brakes

- o Hydraulic and pneumatic power presses

- o Hot bending and hot metal presses

- o Riveting machines and similar types of fastener applicators

☐ <u>Mechanical Power Press Guarding and Construction</u>

- o Hazards to Employees:

    - ◊ Machine components shall be designed, secured, or covered to minimize hazards caused by breakage, or loosening, falling, or release of mechanical energy (i.e., broken springs).

    - ◊ Friction brakes shall be sufficient to stop the motion of the slide quickly and capable of holding the slide and its attachments at any point.

- o Foot Pedals (Treadle)

    - ◊ The pedal mechanism shall be protected to prevent unintended operation from falling or moving objects or by accidental stepping onto the pedal.

    - ◊ The path of travel of pedal counterweights provided shall be enclosed.

- o Hand-operated lever power presses shall be equipped with a spring latch on the operating lever to prevent premature or accidental tripping.

- o Air-controlling equipment shall be protected against foreign material and water entering the pneumatic system of the press.

- o The maximum anticipated working pressure in any hydraulic system on a mechanical power press shall not exceed the safe working pressure rating of any component used in that system.

☐ <u>Point of Operation Safeguarding</u> – The supervisor shall provide and ensure usage of "point of operation guards" or properly applied and adjusted point of operation devices on every operation performed on a mechanical power press.

> **NOTE:** This requirement does not apply when the point of operation opening is 1/4 inch or less.

☐ Point of operation guards guard shall meet the following requirements:

- o Prevent the entrance of hands or fingers into the point of operation.

- o Create no pinch points between the guard and moving machine parts.

- o Conform to the maximum permissible openings (see 1910.217, Table O-10 for more information).

- o Utilize fasteners not readily removable by the operator to minimize the possibility of misuse or removal of essential parts.

- o Facilitate its inspection.

- o Offer maximum visibility of the point of operation.

☐ <u>Point of Operation Devices</u> – Point of operation devices shall protect the operator by:

- o Preventing and/or stopping normal stroking of the press if the operator's hands are inadvertently in the point of operation.

- o Preventing the operator from inadvertently reaching into the point of operation or withdrawing his hands if they are in the point of operation, as the dies close.

- o Requiring both of the operator's hands to operate machine controls during the die closing portion of the press stroke.

- o Locating single-cycle operating controls so the slide completes its downward travel before the operator's hands can inadvertently reach the point of operation.

- o The gate or movable barrier device shall prevent the operator's hands and fingers from entering the point of operation and shall enclose the point of operation before the press clutch can be activated.

- o The presence-sensing point of operation device shall be interlocked into the brake/ clutch control circuit. This will prevent or stop clutch activation and apply the brake if the device in the sensing field detects an operator's hand or other body part.

☐ Dies should allow for the automatic ejection of stock and scrap.

☐ <u>Inspection, Maintenance, and Modification of Presses</u> – The employer shall establish and follow a program of periodic and regular inspections of power presses to ensure that all parts, auxiliary equipment, and safeguards are in safe operating condition and adjustment. The employer shall maintain records of these inspections and the maintenance work performed.

FORGING MACHINES

☐ Use of Lead

- o Thermostatic control of heating elements shall maintain proper melting temperature and prevent overheating.

- o Fixed or permanent lead pot installations shall be exhausted.

- o Portable units shall be used only in areas where there is adequate general room ventilation.

- o Personal protective equipment (gloves, goggles, aprons, and other items) shall be worn.

- o Dross skimmings shall be stored in a covered container.

- o Equipment shall be kept clean, particularly from accumulations of yellow lead oxide.

☐ Inspection, Maintenance – The employer shall maintain all forge shop equipment in a condition to ensure continued safe operation. This includes:

- o Establishing periodic and regular maintenance safety checks and keeping records of these inspections.

- o Scheduling and recording inspections of guards and point of operation protection devices at frequent and regular intervals.

- o Training personnel for the proper inspection and maintenance of forging machinery and equipment.

- o Fastening or protecting all overhead parts so that they will not fly off or fall in the event of a failure.

☐ Presses and Hammers – Hammers shall be positioned or installed on foundations adequate to support them.

- o All presses shall be installed in such a manner that they remain where they are positioned or anchored to foundations sufficient to support them.

- o Means shall be provided for disconnecting the power to the machine and for locking out or rendering cycling controls inoperable.

- o The ram shall be blocked when dies are being changed or other work is being done on the hammer.

- o A substantially constructed scale guard shall be provided at the back of every hammer to stop flying scale.

☐ Die keys and shims shall be made from a grade of material that resists cracking or splintering.

☐ Foot-operated devices (i.e., treadles, pedals, bars, valves, and switches) shall be substantially and effectively protected from unintended operation.

☐ All manually operated valves and switches shall be clearly identified and readily accessible.

☐ Power-Driven Hammers

- o Steam or air hammers shall have a safety cylinder head to act as a cushion if the rod should break or pull out of the ram.

- o Steam or air piping shall conform to the American National Standard ANSI B31.1.0-1967.

☐ <u>Gravity Hammers</u>

    o   Air lift hammers shall:

         ◊   Have a safety cylinder head to act as a cushion if the rod should break or pull out of the ram.

         ◊   Have a conveniently located, distinctly marked air shutoff valve in the admission pipeline.

         ◊   Be provided with two drain cocks: one on the main head cylinder and one on the clamp cylinder.

    o   Air piping shall conform to ANSI B31.1.0-1967, Power Piping with Addenda issued before April 28, 1971.

☐ <u>Board Drop Hammers</u>

    o   Board drop hammers shall have a suitable enclosure to prevent damaged or detached boards from falling. The board enclosure shall be securely fastened to the hammer.

    o   All manually operated valves and switches shall be clearly identified and readily accessible.

TERMS AND DEFINITIONS

☐ ***Calender*** – A process of pressing between plates in order to smooth and glaze or to thin into sheets.

☐ ***Kerf*** – The width of cut made by a saw or cutting torch.

☐ ***Non-kickback fingers or dogs*** – Any of various usually simple mechanical devices for holding, gripping, or fastening that consist of a spike, bar, or hook.

☐ ***Point of operation*** – The area on a machine where work is actually performed upon the material being processed.

☐ ***Rip-saw*** – A coarse-tooth saw used to cut wood in the direction of the grain.

☐ ***Treadle*** – A swivel or lever device pressed by the foot to drive a machine.

----------END OF SECTION----------

Section 23
Material Handling

OVERVIEW

This section provides information regarding the requirements for use of mechanical equipment, storage of materials, housekeeping, clearance limits, rolling railroad cars, and guarding. Specifics for the use and safe operation of powered industrial vehicles such as forklifts or tow units can be found in the Powered Industrial Vehicles Section of this text.

REGULATORY COMPLIANCE

☐ §1910.176 Handling Materials - General

USE OF MECHANICAL EQUIPMENT

☐ Where mechanical equipment is used, safe clearance shall be allowed for loading docks, through doorways, and wherever turns or passage must be made.

☐ Aisles and passageways shall be kept clear and in good repair with no obstruction across or in aisles that could create a hazard.

☐ Permanent aisles or passageways shall be appropriately marked.

SECURE STORAGE

☐ Storage of material shall not create a hazard.

☐ Bags, containers, and bundles shall be stacked, blocked, interlocked, and limited in height so that they are stable and secure against sliding or collapse.

HOUSEKEEPING

☐ Keep storage areas free from accumulation hazards that constitute hazards for tripping, fire, explosion, and pest harborage.

☐ Vegetation should be controlled as necessary.

CLEARANCE LIMITS

☐ Signs must be provided that warn of clearance limits.

ROLLING RAILROAD CARS

☐ Derail and or bumper blocks shall be provided on spur or railroad tracks where a rolling car could contact other cars being worked, entering a building, work, or traffic area.

GUARDING

☐ Covers and/or guardrails shall be provided to protect workers from open vats, pits, tanks, and ditches, etc.

---------- END OF SECTION ----------

Section 24
Medical Services and First Aid

OVERVIEW

This section outlines the minimum requirements for medical services and first aid in today's industrial environment. Employers must evaluate the current operations and the type of accidents that are likely to occur.

REGULATORY COMPLIANCE

☐ §1910.151 Medical Services and First Aid

GENERAL REQUIREMENTS

☐ Employers must ensure the availability of medical personnel for advice and consultation on matters of plant health.

☐ If your company doesn't have an infirmary, clinic, or hospital in close proximity, you must have persons trained to provide first aid.

BP: It is recommended that companies have basic first aid facilities and at least two first aid/CPR trained employees on each shift to protect employees in case of injury.

☐ Employers must have adequate first-aid supplies readily available on all shifts.

BP: It is recommended that companies maintain an AED (Automated External Defibrillator) on site. The provision of an AED requires a prescription and medical direction from a licensed health care provider. This can often be accomplished by such a person on the local fire department staff. You must also provide the fire department and/or emergency services with the location of your unit(s) within the facility. Only those who have successfully completed CPR (cardiopulmonary resuscitation) and AED training may attempt to use the AED.

☐ Quick drenching or flushing facilities are required in work areas where the eyes or body may be exposed to injurious corrosive materials or chemicals.

BP: A common approach that is NOT recommended for employers is the use of hospital emergency rooms for first aid and minor occupational injuries. This practice usually results in overdiagnosis and overactivity for minor incidents. It is strongly recommended that all companies establish a relationship with a certified occupational physician and occupational clinic that understands how to provide the highest quality health care while not unnecessarily making a case "recordable" for OSHA recordkeeping purposes.

RECOMMENDED TRAINING

☐ Basic first aid and CPR training

☐ Automated external defibrillator (AED) training (when providing such equipment)

☐ Bloodborne pathogens training (if required, see the Section on Bloodborne Pathogens within this text)

RECORDKEEPING

☐ Certificates and training materials

☐ Emergency numbers

☐ First aid logs

☐ Inventory checklist for supplies

---------- END OF SECTION ----------

Section 25
Noise (Hearing Conservation)

OVERVIEW

Occupational noise exposure accounted for more than 28,400 OSHA-recordable hearing loss cases in 2004. Excessive occupational noise exposure without proper protection can lead to permanent hearing loss. Because of this, OSHA has developed the Noise Standard that mandates protection against the effects of noise exposure when sound levels exceed those established as permissible exposure limits (PELs).

REGULATORY COMPLIANCE

☐ §1910.95 Occupational Noise Exposure

MONITORING

☐ When employees are subjected to sound exceeding those listed in Table 1 below, feasible engineering or administrative controls shall be utilized.

☐ An effective Hearing Conservation Program shall be administered whenever noise exposures equal or exceed an 8-hour time-weighted average (TWA) of 85 decibels measured on the A scale (slow response) or equivalently, a dose of 50%. This measurement is considered or referred to as the "action level."

☐ Sample work areas suspected of approaching or exceeding the action level to determine what controls (engineering, administrative, or PPE) are most appropriate and to determine inclusion in the hearing conservation program. For employees (even those in low-level noise areas) who are highly mobile within the work area, personal monitoring may be most appropriate.

☐ If controls fail to reduce sound levels within the levels of Table 1, personal protective equipment (PPE) shall be provided and used (required) to reduce the sound levels to within the levels of Table 1. This PPE shall be provided at no cost to the employee.

☐ If sampling indicates an excess of the permissible exposure limit (PEL) as indicated by Table 1, the area monitored shall be included in the (written) Hearing Conservation Program and comply with the requirements of this section. For those areas that do not exceed the PEL, personnel will not be included in the program, but should wear hearing protection when in an affected area.

☐ Monitoring shall be repeated whenever a change in production, process, equipment, or controls increases noise exposures may expose additional employees at or above the action level or may exceed the current attenuation level of hearing protection used by employees.

Table 1 (Permissible Exposure Limits)

The table below outlines the permissible noise exposures or permissible exposure limits (PELs) that should not be exceeded without proper controls:

Duration per Day (Hours)	Decibel Sound Level
8	90
6	92
4	95
3	97
2	100
1 1/2	102
1	105
1/2	110
1/4 or less	115

Notes: 1) When the daily noise exposure is composed of two or more periods of noise exposure of different levels, their combined effect should be considered, rather than individual effect of each. 2) Exposure to impulsive or impact noise should not exceed 140 decibels peak sound pressure level.

☐ The employer shall notify each affected employee at or above the action level as a result of the monitoring.

☐ The employer shall provide affected employees or their representatives with the opportunity to observe any noise measurements conducted.

BP: Although it is required to perform monitoring (or a noise survey) whenever a change in the operation may increase noise levels to affected employees, many businesses will conduct a survey every year or two. Include the results of surveys in the written Hearing Conservation Program by charting (or otherwise specifying) specific areas that were sampled. For sound level meter measurements, provide a small range that quantifies a consistent measurement of the area.

AUDIOMETRIC TESTING

☐ All employees included in the Hearing Conservation Program are required to obtain a baseline audiogram (provided by the employer at no cost to the employee) within 6 months of the employee's initial exposure at or above the action level. Where mobile test vans are used, there is an exception that allows testing within 1 year of the initial exposure. However, employees exposed at or above the action level shall be required to wear hearing protection until the baseline audiogram is obtained.

BP: Although the baseline audiograms are allowed by OSHA after initial exposures, because of workers' compensation liabilities, it is recommended that employers obtain baseline audiograms prior to the initial exposure.

☐ It is imperative that employees avoid high levels of occupational or non-occupational noise exposures during the 14-hour period immediately preceding the audiometric examination.

☐ Following the baseline audiogram, applicable employees will be required to obtain a new audiogram on a yearly basis. The employee's annual audiogram will be compared to the baseline audiogram to determine validity and if a standard threshold shift (STS) has occurred. A STS is defined as a change in the hearing threshold relative to the baseline audiogram of an average of 10dB or more at 2,000, 3,000, and 4,000 hertz in either ear.

☐ An audiologist, otolaryngologist, or a qualified physician must review the audiograms to determine whether there is a need for further evaluation.

☐ If the annual audiogram shows that an employee has suffered a STS, the employer may obtain a retest within 30 days and consider the results of the retest as the annual audiogram.

☐ If the comparison of the annual audiogram to the baseline audiogram indicates a STS, the employee shall be informed in writing within 21 days of the determination.

☐ Unless a physician determines the STS is not work-related, employers shall ensure:
  o Employees not wearing hearing protection are fitted with protectors, trained in their use, and required to wear such protection;
  o Employees already using protection shall be refitted, retrained in their use, and provided protection offering greater attenuation, if necessary;
  o The employee shall be referred to a clinical audiological evaluation or an otological examiniation, as appropriate;
  o And, the employee is informed of the need for an otological examination if a medical pathology of the ear unrelated to the use of hearing protectors is suspected.

☐ An annual audiogram may be substituted for the baseline when the audiologist, otolaryngologist, or physician evaluating the audiogram determines the STS revealed is persistent (has stabilized over multiple subsequent audiograms) or indicates a significant improvement over the baseline.

HEARING PROTECTORS

☐ Employers shall provide a variety of hearing protection for all employees exposed to noise at or above the action level, without cost to the employee.

☐ The hearing protection provided must attenuate noise levels to below a 90dB PEL (85dB PEL for employees who have experienced a threshold shift) as outlined in Table 1.

☐ The employer must require and ensure the proper wearing of hearing protection for those applicable employees. Applicable employees include those exposed to noise at or above the action level, those who (in affected areas) have not received a baseline audiogram, and those who have experienced a standard threshold shift (STS).

TRAINING

☐ All applicable employees, as defined by this section, shall be provided training upon assignment to an applicable exposure location and at least annually thereafter.

☐ Information in the training program shall be updated to include changes in protective equipment and work processes.

☐ Topics to be covered in the Hearing Conservation Training Program will include:
  o The effects of noise on hearing acuity.
  o The purpose of hearing protectors, the advantages and disadvantages of various types, and instructions on selection, fitting, use, and care.
  o The purpose of audiometric testing and an explanation of the test procedures.
  o The company's regulations and enforcement procedures.

☐ **Recording Occupational Hearing Loss on the OSHA 300 Log**

Employers are required to record work-related hearing loss cases when an employee's hearing test showed a marked decrease in overall hearing. If an employee's hearing test (audiogram) reveals a work-related standard threshold shift (STS) for hearing in one or both ears, and the employee's total hearing level is 25 decibels (dB) or more below audiometric zero (averaged at 2,000, 3,000, and 4,000 Hz) in the same ear(s) as the STS, you must record the case on the OSHA 300 Log. Employers can make adjustments for hearing loss caused by aging. Seek the advice of a physician or licensed health care professional to determine if the loss is work-related, and perform additional hearing tests to verify.

☐ **Standard Threshold Shift**:

A standard threshold shift, or STS, is defined in the occupational noise exposure standard at 29 CFR 1910.95(g)(10)(i) as a change in hearing threshold, relative to the baseline audiogram for that employee, of an average of 10 decibels (dB) or more at 2,000, 3,000, and 4,000 hertz (Hz) in one or both ears. In this case the STS must only be reported to the employee (**Note:** These comparisons should only be done by a person at the Technician level or above.). Refer to Example 1 below:

Frequency (Hz)	Baseline (dB)	Current Audiogram (dB)	Difference (dB)
2,000	10	20	10
3,000	5	10	5
4,000	15	30	15
Average	**10**	**20**	**10**

Example 1

☐ **STS & a 25-dB Overall Reduction in Hearing Level**

If the employee has shown an STS, examine against the employee's overall hearing ability in comparison to audiometric zero. Using the employee's current audiogram, average the hearing levels at 2,000, 3,000, and 4,000 Hz to determine whether or not the employee's total hearing loss exceeds 25 dB from audiometric zero. In this case the STS must be reported to the employee **AND** recorded on the OSHA 300 log. (**Note:** These comparisons should only be done by a person at the Technician level or above.). Refer to Example 2 below.

Frequency (Hz)	Baseline (dB)	Current Audiogram (dB)	Difference (dB)
2,000	20	30	10
3,000	30	35	5
4,000	10	25	15
Average	**20**	**30**	**10**

Example 2

☐ **Entering a Hearing Loss Case in the OSHA 300 Log**

Beginning January 1, 2004, a column has been added to the OSHA 300 log that must be checked whenever an occupational noise loss case is recorded.

RECORDKEEPING

☐ Records verifying participation in training and recurrent training as well as audiometric test records will become a permanent part of the employee's file.

☐ Audiometric test records must include the following:

- o Name and job classification of the employee
- o Date of audiogram
- o Examiner's name
- o Date of last acoustic or exhaustive calibration of the audiometer
- o Employee's most recent noise exposure assessment

☐ Audiometric test records shall be maintained for the duration of the affected employee's employment.

☐ A copy of the Standard (1910.95) shall be available to employees and posted in the workplace.

☐ Noise exposure measure records shall be maintained for a minimum of 2 years.

☐ If an employer ceases to do business, required records shall be transferred to the employer's successor.

---------- END OF SECTION ----------

Section 26
Personal Protective Equipment (PPE)

OVERVIEW

The purpose of this section is to outline general requirements for personal protective equipment (PPE) including respirators, eye, head, foot, and fall protection. PPE is not always the best method for controlling hazards. However, it can be the fastest and most economical method of protecting employees from known hazards. Attempts should always be made to eliminate the hazard that warrants PPE.

REGULATORY COMPLIANCE

☐ §1910.132 General Requirements

☐ §1910.133 Eye and Face Protection

☐ §1910.134 Respiratory Protection

☐ §1910.135 Head Protection

☐ §1910.136 Foot Protection

☐ §1910.138 Hand Protection

WRITTEN WORKPLACE HAZARD ASSESSMENT

☐ A written Workplace Hazard Assessment must be conducted to identify potential employee hazards. This is a general assessment of each work activity. For example, for a hypothetical activity of "running a lathe," the hazards to consider would be noise, flying objects, and maybe the potential for foot injury from placing or removing stock. The required PPE in this case would likely be hearing protection, eye and face protection, and foot protection, all of grades necessary to mitigate the hazard. The assessment must be reviewed periodically and revised when new hazards are introduced or altered in the workplace.

☐ The employer must select PPE based on the findings of the Workplace Hazard Assessment to protect employees.

☐ It is important that selected PPE properly fit each employee.

☐ Workplace Hazard Assessments must contain the following information:

    o   Name of the individual performing the assessment;

- o Date of the hazard assessment;
- o The workplace being evaluated;
- o The specific hazards identified during the assessment; and,
- o Certification of the document stating that the evaluation and required information has been completed.

EAR (HEARING) PROTECTION

☐ Hearing protection shall be provided by the employer at no cost to employees exposed at or above an 8-hour time-weighted average (TWA) of 85 decibels. More specifics about ear protection can be found in the Noise (Hearing Conservation) section of this text.

EYE AND FACE PROTECTION

☐ Eye protection is required when there is the possibility of a hazard causing injury or illnesses of the eyes.

☐ Eye protection shall have side shields and be marked with ANSI Z87.1. (Consult a local dealer for details.)

☐ Potential eye and face hazards:
- o Flying objects
- o Glare
- o Liquids
- o Irritant gases
- o Radiation

☐ Protective eyewear shall:
- o Fit comfortably
- o Fit snugly
- o Be shatter resistant
- o Be cleaned and disinfected when necessary

☐ Employees who wear corrective lenses have the following options:
- o Safety glasses with prescription lenses
- o Safety goggles worn over glasses that do not alter the fit of the goggles
- o Goggles with prescription lenses mounted behind the protective lenses of the goggles

HEAD PROTECTION

☐ Head protection must meet the requirements of ANSI Z89.1.

☐ Head protection must be provided to protect the user from falling objects or bumps.

☐ Head protection must fit properly.

☐ Helmets should not be marked or painted unless approved by the manufacturer.

FOOT PROTECTION

☐ Foot protection must meet the requirements of ANSI Z41.1.

☐ Foot protection should be appropriate to the hazard (e.g., steel toes, metatarsal guards, slip resistant).

HAND PROTECTION

☐ Employers shall select and require employees to use appropriate hand protection when exposed to hazards such as those from skin absorption of harmful substances, severe cuts or lacerations, severe abrasions, punctures, chemical burns, and harmful temperature extremes.

☐ Employers shall base selection on performance characteristics such as conditions present, duration of use, and the hazards (or potential hazards) identified.

FALL PROTECTION

☐ Fall protection shall be used when a worker is over 6 feet off the ground.

☐ All fall protection devices shall be inspected for defects before each use.

☐ If fall protection is damaged in any way, it shall not be repaired. Damaged fall protection equipment must be destroyed and thrown out.

☐ If a fall occurs, equipment such as lanyards and harnesses must be replaced.

☐ The following fall protection is currently approved for use:

   o   Full body harnesses.

   o   Lanyards (a tying-off device from a belt or harness to stable structure within reach).

   o   Horizontal lifelines (a cable used when there is nothing convenient to tie off to. It is tied off between two stable structures within reach).

   o   Retractable lifeline (deceleration device which contains a drum-wound line).

☐ Harnesses must have D rings on the front and back for fall protection and rescue purposes.

☐ Harnesses, lifelines, retractable lifelines, and lanyards shall be constructed of a durable material and marked with a tag stating maximum load and name of manufacturer.

☐ As of January 1, 1998, belts are no longer acceptable as fall protection.

RESPIRATORY PROTECTION

☐ The employer shall provide all respiratory equipment to any employee working in areas that require the use of protection. For more specifics on this subject, see the Respiratory Protection section within this text.

TRAINING

☐ The employer shall provide training to each employee who is required by this section to use PPE. The affected employee shall be trained to know:

- o When PPE is necessary

- o What PPE is necessary

- o How to properly don, doff, adjust, and wear PPE

- o The limitations of the PPE

- o The proper care, maintenance, and useful life and disposal of PPE

☐ Each affected employee shall demonstrate an understanding of the training specified in this section and the ability to use PPE properly before being allowed to perform work requiring PPE.

☐ When the employer believes that an affected employee does not have the understanding and skill as required by this section, the employee shall be retrained. Circumstances where retraining is required include, but are not limited to, situations where:

- o Changes in the workplace render the previous training obsolete.

- o Changes in the types of PPE used render previous training obsolete.

- o Inadequacies in an affected employee's knowledge or use of assigned PPE to be used render previous training obsolete.

- o The affected employee has not retained the requisite knowledge or skill for the use of the required PPE.

☐ The employer shall verify each affected employee receives and understands the required training through a written certification that contains the names of each employee trained, the date(s) of the training, and the subject of the certification.

RECORDKEEPING

☐ Written Workplace Hazard Assessment

☐ Inspections as required (daily, monthly, annual).

☐ Training documentation and certifications for the understanding and skill of use of PPE required.

☐ Maintenance and repair records of certain equipment (respiratory).

---------- END OF SECTION ----------

Section 27
Powered Industrial Vehicles

OVERVIEW

This section covers the safety requirements regarding the operation, fire protection, design, and maintenance of motorized material handling and other specialized industrial vehicles. Materials handling vehicles include forklifts, material lift trucks, stock-pickers, electric pallet jacks, motorized hand trucks, and other similar vehicles used to pick up, carry, or otherwise move pallets or palletized material. Specialized vehicles include floor sweepers, tow units, or other motorized equipment with a specific purpose for industrial application.

This section does not address modifications to powered industrial vehicles. Modifications should only be performed in accordance to the manufacturer's specifications and only by a trained or qualified individual (normally a representative of the manufacturer). Modifications include, but are not limited to additions to or removal of originally approved parts of the vehicle.

REGULATORY COMPLIANCE

☐ §1910.178 Powered Industrial Vehicles

GENERAL

☐ Fuel storage and handling shall be conducted in accordance with National Fire Protection Association (NFPA) Regulations. See NFPA 30 (Flammable and Combustible Liquids) or NFPA 58 (Storage and Handling of Liquefied Petroleum Gases), as applicable.

☐ Set brakes and provide wheel stops or chocks for vehicles that must be parked on an incline.

☐ Set brakes and provide wheel stops, chocks, or other recognized positive protection (dock trailer securing devices) while trailers, train cars, etc. are being loaded to prevent rolling while being boarded or loaded.

BATTERY CHARGING AND STORAGE

☐ Battery charging installations shall be located in areas designated for that purpose.

☐ Facilities shall be provided with flushing and neutralizing agents for spilled electrolytes, protection from physical damage, and adequate ventilation for dispersal of fumes from gassing batteries.

☐ A conveyor, overhead hoist, or equivalent material handling equipment shall be provided for handling batteries.

☐ When charging batteries, acid shall be poured into water, not water into acid.

☐ Smoking and/or open flames shall be prohibited in charging areas.

EMPLOYER RESPONSIBILITIES

☐ The employer shall ensure:

- o Operators meet the required qualifications and are certified in accordance with this section.

- o Operators are familiar with the requirements of this section.

- o Operators *never* place their arms or legs between the uprights of the mast or outside the running lines of the truck.

- o A process is in place for observing operators periodically to ensure compliance with the operator responsibilities in this section.

- o A process is established for operators to perform a daily inspection and remove from service any vehicles found unserviceable during inspection. Ensure the vehicle checklists used includes information such as checking oil/transmission fluid levels, leak checks, ANSI markings, load rate markings, structural deformities, etc.

- o Traffic regulations are established, including reinforcing speed limits.

- o Auxiliary directional lighting is provided on the truck where general lighting is not sufficient for the operator to see or less than two lumens per square foot.

- o Forklifts are fitted with an overhead guard unless operating conditions do not permit.

- o Handholds or other effective means are provided on portable dock boards to permit safe handling.

OPERATOR RESPONSIBILITIES/SAFETY GUIDELINES

☐ Vehicle Condition and Maintenance:
Operators shall:

- o Inspect the vehicle daily in accordance with the manufacturer's recommendations.

- o Continue to monitor the condition of the vehicle throughout the shift and immediately report any new deficiencies or concerns to supervision, while also initiating the process for having the deficiency repaired.

- o Report to supervisor any incidents of damage immediately.

CAUTION: No person shall operate a vehicle that has been tagged out of service!

- o Remove from service any vehicle if the deficiency warrants it by the manufacturer's recommendation or the vehicles' inspection. Identify the vehicle with a tag stating "Out of Service" or similar verbiage so that no one else will use the vehicle inadvertently.

☐ Loads and Loading:
Operators shall:

- o *Capacity:*

  - ◊ Be familiar with the rated capacity of the vehicle.

  - ◊ *Never* lift or transport loads that exceed the rated capacity of the vehicle.

- o *Visibility:*

  - ◊ Not lift or transport loads where visibility may be obstructed.

  - ◊ Operate the vehicle in reverse when carrying loads that restrict forward visibility.

◊   Be mindful of the dimensions of the load in front when operating in reverse.

◊   Use a spotter or guide person when driving forward with an obstructed view is absolutely necessary (driving into trailer, up grades, etc.).

◊   Drive almost at idle speed when using a spotter.

o   *Loading:*

◊   Handle only loads that are securely and uniformly stacked or otherwise stable.

◊   Use caution when handling loads that cannot be centered.

◊   Carry loads that fall within the rated capacity of the vehicle.

◊   Ensure the load is fully engaged from the bottom and then tilt the mast backward to stabilize the load before moving.

◊   Only tilt an elevated load forward when the load is in a deposit position over a rack or stack.

◊   *Never* use damaged pallets.

o   *Large Loads:*

◊   Use a vehicle with a vertical load backrest extension when handling loads that are high or segmented.

◊   Keep the load as low as possible so that the center of gravity is as low as possible. (This will maximize the vehicle's capacity.)

◊   Move long, high, or wide loads slowly, carefully, and in a manner that they are not pulled, pushed, or dragged.

◊   *Never* use personnel or counterweights to increase the capacity of the vehicle. Carry the load with the blades close to the ground and mast tilted backward.

☐   Know Surrounding Operations:

The following addresses the requirements regarding operations that are potentially occurring around or "nearby" while operating a forklift or similar vehicle:

Operators shall:

o   *Sound Horn*:

◊   Stop and sound the horn before proceeding at all cross-aisles, corners, or other areas with limited visibility. (This will alert others approaching the same area to your presence.)

o   *Facility*:

◊   Maintain at least 10 feet from any overhead electrical wires.

◊   Maintain clearance between other overhead obstructions (pipes, low doors, etc.) as needed to prevent contact or damage.

◊   *Never* use a vehicle for opening or closing freight doors or truck doors.

o   *Traffic Precautions*:

◊   Watch for pedestrian traffic *at all times* and yield them the right-of-way.

◊   Ensure *no one at any time* stands or passes under elevated loads, empty forks, attachments, or any other elevated portion of the vehicle.

◊   Maintain at least three vehicle lengths between vehicles when following in the same direction.

◊ *Never* pass other vehicles traveling in the same direction at intersections, blind spots, or other hazardous locations.

☐ Floor and Other Driving Surfaces:
Operators shall:

o *Proceed Cautiously*:

◊ Keep a safe distance (at least one vehicle length) from the edge of ramps, platforms, or floor openings at all times.

◊ Slow down and use extra caution when operating on wet or slippery floors.

◊ Avoid running over loose objects, bumps, holes, or other irregular surfaces.

◊ Cross rails or tracks in the surface at a diagonal angle whenever possible.

o *Inclined Surfaces*:

◊ *Never* use a forklift on a grade higher than a 5% incline. Keep the load "upgrade" when ascending or descending on an inclined surface, and the forks "downgrade" when not handling a load.

◊ Have the load tilted back and raised only as high as necessary to clear the surface when operating on an inclined surface.

◊ Keep the forks downgrade and in the raised position when using a pallet jack on a grade.

◊ *Never* ride a pallet jack on a grade.

o *Dock Boards (Bridge Plates)*:

◊ Secure dock boards by anchoring or using other suitable means so that they will not move when driving over them.

◊ Drive over dock boards slowly; do not exceed their rated capacity.

◊ Check dock boards for handholds and report deficiencies (**BP**: to your Facilities Maintenance Department).

o *Tractor Trailers*:

◊ Verify the brakes of tractor-trailer trucks are set and wheel chocks placed under the rear wheels to prevent the trucks from rolling while they are being boarded.

NOTE: A positive mechanical means of securing trailers to the loading dock may be used in place of the chock requirement providing the system is installed and used in a manner that effectively prevents movement of the trailer throughout the boarding, loading, and unloading operation.

☐ Driving or Operator Limitations:
Operators Shall:

o *General*:

◊ Use auxiliary directional lighting on the truck where general lighting is less than two lumens per square foot.

o *Driving*:

◊ Observe all traffic regulations, including authorized area speed limits.

◊ Travel on the right side of the aisle or roadway whenever possible.

◊ Report all incidents with personnel, the facility, structures, etc., immediately, regardless of whether injury or damage is apparent.

◊ Include the names and addresses of any witnesses in the report.

◊ Lower the forks, put the transmission in neutral and set the parking brake *whenever leaving the operator's seat for any reason.*

◊ Comply with the requirements following this subsection when parking, leaving, or walking out of the sight of the vehicle.

o *Speed*:

◊ Under *all travel conditions* and *at all times*, the vehicle shall be operated at a speed that will permit it to be brought to a stop in a safe manner (no faster than a brisk walk while carrying a load).

◊ Do not drive too fast for the environmental and surrounding conditions.

◊ Slow down before entering a turn—especially when the forks are raised for load clearance.

◊ Do not race the engine to speed the lifting or tilting action of forklifts.

o *Personnel*:

◊ Ensure *no one* stands or passes under elevated loads, empty forks attachments, or any other elevated portion of the vehicle.

◊ *Never* drive up to anyone standing in front of a bench or other fixed object so that if the truck failed to stop, the individual standing in front would be pinned against the object.

◊ *Never* place arms or legs between the uprights of the mast or outside the running lines of the truck.

◊ *Never* allow someone to ride on any vehicle or equipment towed by vehicles without their occupying a seat or other equipment designed for a person to ride.

◊ *Never* engage in horseplay of any kind.

☐ Parking or Leaving Unattended (more than 25 feet away or out of visual contact): Operators shall:

o Ensure vehicles are parked in such a manner that they do not obstruct aisles, exits, access to stairways, or safety equipment.

o Position in such a way that other traffic will not be obstructed.

o Position on level ground whenever possible.

o Chock at least one wheel when it is necessary to park on an incline.

o Fully lower the load-engaging means (forks, etc.).

o Neutralize the controls (transmission to neutral, lift controls neutral, etc.).

o Shut off the power (stop the engine, etc.).

o Set the parking brake.

TRAINING AND QUALIFICATIONS

☐ Trainer / Evaluator Qualifications:

Instructors (trainers/evaluators) shall be individuals who have the necessary knowledge and experience to ensure that operators gain the skills needed to operate the vehicle in an acceptable manner. Their qualifications may include professional standing, a recognized degree, or demonstrated ability through extensive knowledge, training, and experience.

☐ Individuals:

To be certified as an operator, individuals shall:

o   Be at least 18 years of age.

o   Possess a valid state or international driver's license.

NOTE: Loss or suspension of a state or international license disqualifies the individual as an operator and requires the individual to notify supervision and refrain from operating company vehicles immediately. Failure to do so will invoke disciplinary action against the individual up to and including termination.

---

o   Complete initial training for the type of equipment before operating.

o   Complete a re-certification evaluation for the specific piece of equipment at least once every 3 years.

o   Complete refresher training when any one of the following applies:

◊   Operator is observed operating the vehicle in an unsafe manner.

◊   Operator is involved in an accident or near-miss incident.

◊   Operator has an evaluation that reveals he/she is not operating the vehicle safely.

◊   Operator is assigned to drive a different type of vehicle, or

◊   A condition in the work area changes in a manner that could affect safe operation of the vehicle.

INITIAL TRAINING

The employer shall ensure each operator complies with this subsection *for each type* of vehicle he/she is expected to operate *before* he/she is allowed to operate the vehicle:

☐ Requirements:

Certification requires all three of the following for each vehicle:

o   <u>Classroom Training</u>:

Formal instruction consisting of *at least one* of the following: lecture, discussion, interactive computer learning, videotape, or written material.

o   <u>Practical</u>:

Hands-on or on-job training (OJT) consisting of *both* demonstrations performed by the trainer and practical exercises performed by the trainee. This shall be performed on the type of equipment for which he/she is being certified in the workplace in which he/she will operate.

NOTE: The trainee shall only operate the vehicle under the direct supervision of the trainer and only where such operation will not endanger the trainee or other employees.

---

o   <u>Evaluation</u>:

An evaluation conducted on the trainee's actual performance (driving test) that verifies his/her ability to control the vehicle. This shall be conducted with the type of equipment for which he/she is being certified in the workplace in which he/she will operate.

☐ Program Content:

Certification requires that all of the following topics be covered in the formal, practical, and evaluation phases of training for the type of vehicle on which the operator is being certified. (The only exception is in cases where the topics are not applicable to the operable performance of the vehicle in the operation.)

- o Vehicle-Specific:

  ◊ Operating instructions, warnings, and precautions for the vehicle.

  ◊ Differences between the controls and handling characteristics of the vehicle and a regular automobile.

  ◊ Vehicle controls and instrumentation: where they are located, what they do, and how they work.

  ◊ Operating characteristics including steering, maneuvering, visibility (including restrictions or blind spots), capacity, stability, and operating restrictions.

  ◊ Attachments or other devices including forks, drum carriers, etc. along with their use limitations.

  ◊ Vehicle-specific inspection items and procedures along with any maintenance the operator is required to perform.

  ◊ Battery recharging procedures and precautions.

  ◊ Any other operating instructions, warnings, or precautions listed in the operator's manual for the vehicle.

- o Area-Specific:

  ◊ Surface conditions under which the vehicle is operated.

  ◊ Ramps and other sloped surfaces that could affect the vehicle's stability.

  ◊ Pedestrian and vehicle traffic in the area where the vehicle is operated.

  ◊ Narrow aisles (indoor or out) and other restricted or congested areas where the vehicle is operated.

  ◊ Other unique or potentially hazardous conditions in the area of operations that could affect safe operation.

- o Load-Carrying Vehicles:

  ◊ Composition of loads to be carried and load stability

  ◊ Load manipulation, stacking, and unstacking

REFRESHER TRAINING

The employer shall require that operators receive refresher training when called for by the Training and Qualifications Subsection above.

☐ Refresher training shall include either all the elements of initial training or specific content targeted at the specific element that required the operator's need for refresher training (e.g., targeted at the new vehicle/condition or the operator's undesired behavior, etc.).

BP: Refresher training may also consist of documented employee counseling by a manager, supervisor, or safety representative, where applicable.

☐ Refresher training also will include an evaluation of the effectiveness of the refresher training itself to ensure that the operator has the knowledge and skills needed to operate the vehicle properly **before** he/she is re-certified.

RE-CERTIFICATION

The employer shall require operators to be evaluated for re-certification at least once every 3 years.

☐ Eligibility Verification:

Operators shall verify they hold a valid state/international driver's license by presenting the license to the evaluator upon request.

☐ Requirements:

o Operators shall demonstrate through actual performance that they have the knowledge and ability to control the vehicle.

o Qualified evaluators (See Trainer/Evaluator Qualifications in the Training and Qualifications Subsection) can fulfill re-certification requirements through a visual verification that the operator complies with the safety guidelines of this Section.

o The evaluator may also ask the operator questions pertaining to his knowledge of safety practices regarding operation of the equipment.

o If the above requirements are met satisfactorily, the evaluator shall certify that the operator can safely continue to operate the equipment through another triennial period. Otherwise, the evaluator shall recommend further training and/or exclude the operator from use of the equipment.

REPORTING AND RECORDKEEPING

☐ Reporting

o Operators shall report loss or suspension of their state/international license to the employer.

o Operators shall immediately report incidents involving equipment that result in employee injury or damage to equipment or other property.

☐ Recordkeeping

o The employer shall maintain a record of their operator's initial qualification or re-certification information on file for at least 3 years or until the operator's next evaluation.

o Daily inspections must be maintained (**BP**: for at least the previous 30 days).

---------- END OF SECTION ----------

Section 28
Recordkeeping and Reporting of Occupational Injuries and Illnesses

---

OVERVIEW

This section provides information regarding the requirements to record and report work-related fatalities, injuries, and illnesses. Remember, simply recording a work-related injury, illness, or fatality does not mean that the employer or employee was at fault. It also doesn't mean that an OSHA rule has been violated or that the employee is eligible for worker's compensation or other benefits. ALL EMPLOYERS regardless of size, industry, or exemptions status MUST report to OSHA any workplace incident that results in a fatality or the hospitalization of three (3) or more employees within eight (8) hours.

NOTE: OSHA provides a 12-page printable guideline at www.osha.gov to help employers with completing forms.

REGULATORY COMPLIANCE

☐ Title 29 Code of Federal Regulations (CFR) 1904

PARTIAL EXEMPTIONS FOR SIZE AND INDUSTRY

☐ If your company had ten (10) or fewer employees at all times during the previous calendar year, you do not need to keep OSHA records.

☐ However, if you have ten (10) or fewer employees and you are informed in writing by OSHA or the BLS (Bureau for Labor Statistics) that you must complete an annual survey, it must be completed accurately and returned promptly.

☐ If you had more than ten (10) employees **at any time** during the last calendar year, you must keep OSHA injury and illness records unless you are in a partially exempt industry.

o   Examples of partially exempt (low hazard) industries are retail, service, finance, real estate, or insurance.

o   Even companies in partially exempt industries must complete surveys received from OSHA or the BLS.

o   If you own multiple companies, you must keep OSHA injury and illness records for all non-exempt establishments unless you are partially exempted because of size.

RECORDING CRITERIA

Employers must keep records of fatalities, injuries, and injuries that are:

☐ Work-related; and

☐ A new case; and

☐ Meets one or more of the following criteria:
  o Meets OSHA standards to be classified as recordable:
    ◊ Death
    ◊ Days away from work
    ◊ Restricted work or transfer to another job
    ◊ Medical treatment <u>beyond</u> first aid. First aid (and NOT recordable) includes:
      ☐ Non-prescription medication at non-prescription strength
      ☐ Tetanus immunizations or hepatitis B vaccines
      ☐ Cleaning, flushing, or soaking wounds on the surface of the skin
      ☐ Using wound coverings such as bandages, gauze pads, or butterfly bandages
      ☐ Using hot or cold therapy
      ☐ Using any non-rigid means of support such as elastic bandages, wraps, non-rigid back belts
      ☐ Using temporary immobilization devices while transporting a victim
      ☐ Drilling a fingernail or toenail to relieve pressure or drain fluid from a blister
      ☐ Using eye patches
      ☐ Removing foreign objects from the eye using only irrigation or a cotton swab
      ☐ Removing splinters or foreign material from areas other than the eye by irrigation, tweezers, cotton swabs, or other simple means
      ☐ Using finger guards
      ☐ Using massages (physical therapy or chiropractic treatment are considered medical treatment and are recordable)
      ☐ Drinking fluids for relief of heat stress
  o A needle stick or cut from any sharp objects that are contaminated with another person's blood or potentially infected body fluids.
  o A worker is medically restricted to perform work such as lifting or walking.
  o A worker misses scheduled days from work starting with the day after the incident.
  o A hearing loss that is considered a work-related standard threshold shift (STS) in one or both ears, and the employee's total hearing level is 25 decibels (dB) or more above the audiometric zero (averaged at the 2,000, 3,000, 4,000 Hz) in the same ear(s).

  BP: Always confirm the hearing loss is worked-related by sending the employee to a medical specialist who can determine the loss was not caused by factors such age or other health problems.
  o A worker is occupationally (performing work duties such as a nurse or medical technician) exposed to anyone with a known case of active tuberculosis (TB) and that

employee later develops TB that is diagnosed by a physician or other licensed health care professional. Record the case and check the "respiratory condition" column.

- o A worker develops a work-related musculoskeletal disorder (MSD) such as, but not limited to carpal tunnel syndrome, rotator cuff syndrome, sciatica, tendonitis, and low back pain.

MULTIPLE BUSINESS ESTABLISHMENTS

☐ Employers must keep a separate OSHA 300 Log for each establishment for each establishment that is expected to be in operation for 1 year or longer.

☐ Employers may keep a single OSHA 300 Log that covers all short-term (will exist less than 1 year) establishments. These records may be kept at your headquarters or other central location if:

- o You can transmit information to the central location to record the injuries or illnesses within seven (7) days; and

- o You can produce and send the records to the other locations as needed by a government representative, employees, former employees, or employee representatives.

☐ If an employee of one establishment is injured at another location, you must record the injury and illness on the OSHA 300 Log of the establishment at which the injury and illness occurred.

COVERED EMPLOYEES

☐ Employers must record on the OSHA 300 Log the recordable injuries and illness of ALL employee on your payroll, whether they are labor, executive, hourly, part-time, seasonal, or migrant workers.

☐ You must also record the recordable injuries and illnesses that occur to employees who are not on your payroll if you supervise these employees on a day-to-day basis. (NOTE: This includes any temporary or placement agency employees that are supervised by your company at your location.)

☐ You don't have to record the injuries and illness of employees who are owners or partners in a sole proprietorship or partnership.

ANNUAL SUMMARY OSHA 300A

☐ At the end of each calendar year, employers must:

- o Review the OSHA 300 Log to verify that entries are complete and accurate.

- o Correct any deficiencies noted on the summary.

- o Post the annual summary (300A) in a conspicuous place or places where notices to employees are usually posted. Make sure the 300A is not altered, defaced, or covered by other material.

- o The summary (300A) must be posted no later than February 1 of the year following the year covered by the records and remain posted until April 30.

RETENTION AND UPDATING

☐ Employers must save the OSHA 300 Log, 300A and all 301 forms for the **5 years** following the end of the calendar year that these records cover.

- [ ] Stored OSHA 300 Logs must be updated or changed during the 5-year period as cases change. Changes include:
    - o Changes in original outcomes such as determination that the incident was NOT work related.
    - o You may simply "line out" the original entry and enter a new entry on the Log.

- [ ] Stored OSHA 300A summaries or 301 Incident Reports need not be updated during the storage period.

CHANGES IN BUSINESS OWNERSHIP

- [ ] If your business changes ownership, you are responsible for recording and reporting work-related injuries and illnesses for the period of the year that you owned the business.

- [ ] The new owner must save old records for the remainder of year and the 5-year storage period but need not update or correct the records of the previous owner.

EMPLOYEE INVOLVEMENT

- [ ] The employer must inform an employee of how they are to report an injury or illness.

- [ ] The employer must provide limited access to injury and illness records to employees and their representatives. The first copy must be provided at no charge. Subsequent copies may incur a reasonable charge.

- [ ] Names on the OSHA Logs must be left intact unless it is a "privacy concern case" such as:
    - o Intimate or reproductive body parts
    - o Injuries from sexual assault
    - o Mental Illnesses
    - o HIV, hepatitis, or tuberculosis
    - o Needle sticks
    - o If the injured employee independently or voluntarily requests privacy

Figure 1: OSHA 300 Log of Work-Related Injuries and Illnesses

OSHA's Form 300 (Rev. 01/2004)

Log of Work-Related Injuries and Illnesses

Attention: This form contains information relating to employee health and must be used in a manner that protects the confidentiality of employees to the extent possible while the information is being used for occupational safety and health purposes.

Year _____

U.S. Department of Labor
Occupational Safety and Health Administration

Form approved OMB no. 1218-0176

You must record information about every work-related injury or illness that involves loss of consciousness, restricted work activity or job transfer, days away from work, or medical treatment beyond first aid. You must also record significant work-related injuries and illnesses that are diagnosed by a physician or licensed health care professional. You must also record work-related injuries and illnesses that meet any of the specific recording criteria listed in 29 CFR 1904.8 through 1904.12. Feel free to use two lines for a single case if you need to. You must complete an injury and illness incident report (OSHA Form 301) or equivalent form for each injury or illness recorded on this form. If you're not sure whether a case is recordable, call your local OSHA office for help.

Establishment name _____

City _____ State _____

Identify the person

(A) Case No.	(B) Employee's Name	(C) Job Title (e.g., Welder)	(D) Date of injury or onset of illness (mo./day)

Describe the case

(E) Where the event occurred (e.g. Loading dock north end)	(F) Describe injury or illness, parts of body affected, and object/substance that directly injured or made person ill (e.g. Second degree burns on right forearm from acetylene torch)

Classify the case

CHECK ONLY ONE box for each case based on the most serious outcome for that case:

Death (G)	Days away from work (H)	Remained at work — Job transfer or restriction (I)	Remained at work — Other recordable cases (J)

Enter the number of days the injured or ill worker was:

Away From Work (days) (K)	On job transfer or restriction (days) (L)

Check the "injury" column or choose one type of illness:

(M)

Injury (1)	Skin Disorder (2)	Respiratory Condition (3)	Poisoning (4)	Hearing Loss (5)	All other illnesses (6)

Page totals | 0 | 0 | 0 | 0 | 0 | 0 | 0 | 0 | 0 | 0 | 0 |

Be sure to transfer these totals to the Summary page (Form 300A) before you post it.

| | (1) Injury | (2) Skin Disorder | (3) Respiratory Condition | (4) Poisoning | (5) Hearing Loss | (6) All other illnesses |

Public reporting burden for this collection of information is estimated to average 14 minutes per response, including time to review the instruction, search and gather the data needed, and complete and review the collection of information. Persons are not required to respond to the collection of information unless it displays a currently valid OMB control number. If you have any comments about these estimates or any aspects of this data collection, contact: US Department of Labor, OSHA Office of Statistics, Room N-3644, 200 Constitution Ave, NW, Washington, DC 20210. Do not send the completed forms to this office.

Page _____ 1 of 1

Figure 2: OSHA 300-A Summary of Work-Related Injuries and Illnesses

OSHA's Form 300A (Rev. 01/2004)

Summary of Work-Related Injuries and Illnesses

All establishments covered by Part 1904 must complete this Summary page, even if no injuries or illnesses occurred during the year. Remember to review the Log to verify that the entries are complete

Using the Log, count the individual entries you made for each category. Then write the totals below, making sure you've added the entries from every page of the log. If you had no cases write "0."

Employees former employees, and their representatives have the right to review the OSHA Form 300 in its entirety. They also have limited access to the OSHA Form 301 or its equivalent. See 29 CFR 1904.35, in OSHA's Recordkeeping rule, for further details on the access provisions for these forms.

Year _____

U.S. Department of Labor

Occupational Safety and Health Administration

Form approved OMB no. 1218-0176

Number of Cases

Total number of deaths	Total number of cases with days away from work	Total number of cases with job transfer or restriction	Total number of other recordable cases
0	0	0	0
(G)	(H)	(I)	(J)

Number of Days

Total number of days away from work	Total number of days of job transfer or restriction
0	0
(K)	(L)

Injury and Illness Types

Total number of...

(M)

(1) Injury — 0
(2) Skin Disorder — 0
(3) Respiratory Condition — 0
(4) Poisoning — 0
(5) Hearing Loss — 0
(6) All Other Illnesses — 0

Post this Summary page from February 1 to April 30 of the year following the year covered by the form

Public reporting burden for this collection of information is estimated to average 50 minutes per response, including time to review the instruction, search and gather the data needed, and complete and review the collection of information. Persons are not required to respond to the collection of information unless it displays a currently valid OMB control number. If you have any comments about these estimates or any aspects of this data collection, contact: US Department of Labor, OSHA Office of Statistics, Room N-3644, 200 Constitution Ave. NW, Washington, DC 20210. Do not send the completed forms to this office.

Establishment Information

Your establishment name _____

Street _____

City _____ State _____ Zip _____

Industry description (e.g., Manufacture of motor truck trailers) _____

Standard Industrial Classification (SIC), if known (e.g., SIC 3715) _____

OR North American Industrial Classification (NAICS), if known (e.g., 336212) _____

Employment Information

Annual average number of employees _____

Total hours worked by all employees last year _____

Sign here

Knowingly falsifying this document may result in a fine.

I certify that I have examined this document and that to the best of my knowledge the entries are true, accurate, and complete.

Company executive _____ Title _____

Phone _____ Date _____

Figure 3: OSHA 301 First Report of Occupational Injury and Illness Form

OSHA's Form 301
Injuries and Illnesses Incident Report

U.S. Department of Labor

Occupational Safety and Health Administration

Form approved OMB no. 1218-0176

Attention: This form contains information relating to employee health and must be used in a manner that protects the confidentiality of employees to the extent possible while the information is being used for occupational safety and health purposes.

This *Injury and Illness Incident Report* is one of the first forms you must fill out when a recordable work-related injury or illness has occurred. Together with the *Log of Work-Related Injuries and Illnesses* and the accompanying *Summary*, these forms help the employer and OSHA develop a picture of the extent and severity of work-related incidents.

Within 7 calendar days after you receive information that a recordable work-related injury or illness has occurred, you must fill out this form or an equivalent. Some state workers' compensation, insurance, or other reports may be acceptable substitutes. To be considered an equivalent form, any substitute must contain all the information asked for on this form.

According to Public Law 91-596 and 29 CFR 1904, OSHA's recordkeeping rule, you must keep this form on file for 5 years following the year to which it pertains.

If you need additional copies of this form, you may photocopy and use as many as you need.

Information about the employee

1) Full Name _____

2) Street _____
City _____ State ____ Zip ____

3) Date of birth _____

4) Date hired _____

5) ☐ Male
☐ Female

Information about the physician or other health care professional

6) Name of physician or other health care professional
_____

7) If treatment was given away from the worksite, where was it given?
Facility _____
Street _____
City _____ State ____ Zip ____

8) Was employee treated in an emergency room?
☐ Yes
☐ No

9) Was employee hospitalized overnight as an in-patient?
☐ Yes
☐ No

Information about the case

10) Case number from the Log _____ (Transfer the case number from the Log after you record the case.)

11) Date of injury or illness _____

12) Time employee began work _____ AM/PM

13) Time of event _____ AM/PM ☐ Check if time cannot be determined

14) What was the employee doing just before the incident occurred? Describe the activity, as well as the tools, equipment or material the employee was using. Be specific. Examples: "climbing a ladder while carrying roofing materials", "spraying chlorine from hand sprayer", "daily computer key-entry."

15) What happened? Tell us how the injury occurred. Examples: "When ladder slipped on wet floor, worker fell 20 feet", "Worker was sprayed with chlorine when gasket broke during replacement", "Worker developed soreness in wrist over time."

16) What was the injury or illness? Tell us the part of the body that was affected and how it was affected; be more specific than "hurt", "pain", or "sore." Examples: "strained back", "chemical burn, hand", "carpal tunnel syndrome."

17) What object or substance directly harmed the employee? Examples: "concrete floor", "chlorine", "radial arm saw." If this question does not apply to the incident, leave it blank.

18) If the employee died, when did death occur? Date of death _____

Completed by _____
Title _____
Phone _____ Date _____

Public reporting burden for this collection of information is estimated to average 22 minutes per response, including time for reviewing instructions, searching existing data sources, gathering and maintaining the data needed, and completing and reviewing the collection of information. Persons are not required to respond to the collection of information unless it displays a current valid OMB control number. If you have any comments about this estimate or any other aspects of this data collection, including suggestions for reducing this burden, contact: US Department of Labor, OSHA Office of Statistics, Room N-3644, 200 Constitution Ave. NW, Washington, DC 20210. Do not send the completed forms to this office.

---------- END OF SECTION ----------

Section 29
Respiratory Protection

OVERVIEW

The purpose of this section is to establish guidelines for the selection, use, care, and maintenance of respirators and to comply with the regulatory requirements of OSHA's 29 CFR 1910.134, Respiratory Protection Standard. This section applies to all employees who work in atmospheres that require the use of respirators. The job task at hand, the products or chemicals being used, and the location of the task determine hazardous work atmospheres.

The employer is obligated to provide a workplace free of airborne contaminants where possible. To do so, the employer should incorporate engineering controls where feasible to eliminate hazards. However, there may be some operations where engineering controls have been employed to minimize exposure, but cannot always eliminate the contaminants. Therefore, the employer's obligation is to further protect employees in these areas through the use of personal protective equipment.

On August 24, 2006, OSHA published in the *Federal Register* new Assigned Protection Factors (APFs) that will complete the 1998 revisions to the Respiratory Protection Standard. APFs are numbers that indicate the level of workplace respiratory protection that a respirator or class of respirators is expected to provide to employees when used as part of an effective respiratory protection program. An APF table is being included in the final standard to guide employers in the selection of air-purifying, powered air-purifying, supplied-air (or airline respirator), and self-contained breathing apparatus (SCBA) respirators. This information and the APF table has been included in this section.

REGULATORY COMPLIANCE

☐ §1910.134 Respiratory Protection

TERMS/DEFINITIONS

☐ *Air purifying respirator* means a respirator with an air purifying filter, cartridge, or canister that removes specific air contaminants by passing the air through an air-purifying element.

☐ *Atmosphere (air) supplying respirator* means a respirator that supplies the user with breathing air from a source such as a tank. Examples are Supplied Air Respirators (SARs) and Self Contained Breathing Apparatus (SCBA) units.

☐ *Canister or cartridge* means a container with a filter, sorbent, or catalyst, or combination of these items, which removes the contaminant from the air passing through the container.

☐ *Employee exposure* means exposure to airborne contaminants that would not occur if the employee were wearing a respirator.

☐ *End of Service Life Indicator (ESLI)* means a system that warns respirator users of the approaching end of adequate protection from a respirator. Respirator cartridges can saturate with contaminants and lose effectiveness.

☐ *Escape-only respirator* means a respirator intended for use only in emergency exit.

☐ *Filter or air-purifying element* means a component that removes solids, liquids, or aerosols from the air.

☐ *Filtering facepiece (dust mask)* means a negative pressure particulate respirator with a filter or respirator in which the entire facepiece is the filter.

☐ *Fit test* means a procedure to ensure a respirator adequately fits the user

☐ *Immediately Dangerous to Life or Health (IDLH)* means an atmosphere that would threaten life, cause irreversible adverse health effects, or impair an individual's ability to escape from a dangerous atmosphere.

☐ *Interior structural firefighting* means the physical activity of fire suppression, rescue or both, inside of buildings or enclosed structures, which are involved in a fire situation beyond the incipient stage. Incipient stage fires are the earliest stages of a small fire where the fire has not spread beyond the original materials of ignition.

☐ *Loose-fitting facepiece* means a respiratory inlet covering that is designed to form a partial seal with the face. The facepiece of a SCBA is an example.

☐ *Negative-pressure respirator (tight-fitting)* means a respirator that is sealed by the negative pressure created during inhalation of the user.

☐ *NIOSH* is the National Institute for Occupational Safety and Health. This is the federal agency responsible for conducting research and making recommendations for the prevention of work-related injury and illness. NIOSH is part of the Centers for Disease Control and Prevention (CDC) in the Department of Health and Human Services.

☐ *Oxygen-deficient atmosphere* means an atmosphere with oxygen content below 19.5% by volume.

☐ *Physician or Other Licensed Health Care Professional (PLHCP)* means an individual who is licensed, registered, or certified by health care standards to provide health care to the public. Physicians, physician assistants, and nurses are examples of PLHCP.

☐ *Positive-pressure respirator* means a respirator in which the pressure inside the facepiece exceeds ambient air outside the respirator.

☐ *Powered air-purifying respirator* means a respirator that uses a blower to force ambient air through the filters to the inside of the respirator facepiece.

☐ *Pressure-demand respirator* means a respirator that admits air into the facepiece when the wearer inhales.

☐ *Qualitative Fit Test (QLFT)* means a pass-fail fit test to determine the fit of a respirator.

☐ **_Quantitative Fit Test (QNFT)_** means an assessment of the adequacy of the respirator by numerically measuring the amount of leakage into the respirator.

☐ **_Self-Contained Breathing Apparatus (SCBA)_** means an air-supplying respirator for which the user carries the tank of air.

☐ **_Service life_** means the period of time a respirator, filter, or sorbent, or other respiratory equipment provides adequate protection to the user.

☐ **_Supplied air or air-line respirator_** means an air-supplying respirator for which the source of breathing air is not designed to be carried by the user.

☐ **_Tight-fitting facepiece_** means a respirator inlet covering that forms a complete seal with the face.

☐ **_User seal check_** means an action conducted by the respirator wearer to determine if the respirator is properly sealed to the face.

TRAINING/QUALIFICATIONS

☐ Personnel using respirators shall be taught the proper selection, use, limitations, and maintenance of respirators, as detailed in this section. In addition, they shall receive instructions in the following:

  o The nature of the hazards in their area, why they must wear a respirator, and what may happen if they neglect to use or improperly use the respirator.

  o The engineering controls being used in their area to minimize the need for respiratory protection.

  o The reasons for selection of a particular type of respirator.

  o The limitations of the selected respirator.

> **NOTE:** This training will be conducted and documented concurrent with the fit test procedure.

RESPIRATOR SELECTION

☐ Respirators will be selected based on respiratory hazard(s) to which the worker is exposed as well as workplace and user factors that affect the respirator performance and reliability.

☐ A new revision to the standard has been made to help employers and employees select the right respirator for the job. Assigned Protection Factors or APFs are numbers that indicate the level of workplace respiratory protection that a respirator or class of respirators is expected to provide to employees when used as part of an effective respiratory protection program. An APF table is being included in the final standard to guide employers in the selection of air-purifying, powered air-purifying, supplied-air (or airline respirator), and self-contained breathing apparatus (SCBA) respirators. (See Table 1 below.)

☐ Employers must follow these new requirements and use APFs to select the appropriate type of respirator based upon the exposure limit of a contaminant and the level of the contaminant in the workplace. Employers select respirators by comparing the exposure level found in the workplace and the maximum concentration of the contaminant in which a particular type of respirator can be used (the Maximum Use Concentration, or MUC). Employers generally determine the MUC by multiplying the respirator's APF by the contaminant's exposure limit. If the workplace level of the contaminant is expected to exceed the respirator's MUC, the employer must choose a respirator with a higher APF.

Table 1

Assigned Protection Factors[5]

Type of Respirator[1,2]	Quarter Mask	Half Mask	Full Facepiece	Helmet/Hood	Loose-Fitting Facepiece
1. Air-Purifying Respirator	5	10[3]	50	—	—
2. Powered Air-Purifying Respirator (PAPR)	—	50	1,000	25/1,000[4]	25
3. Supplied-Air Respirator (SAR) or Airline Respirator • Demand mode • Continuous flow mode • Pressure-demand or other positive-pressure mode	— — —	10 50 50	50 1,000 1,000	25/1,000[4]	— 25 —
4. Self-Contained Breathing Apparatus (SCBA) • Demand mode • Pressure-demand or other positive-pressure mode (e.g., open/closed circuit)	— —	10 —	50 10,000	50 10,000	— —

Notes:

[1] Employers may select respirators assigned for use in higher workplace concentrations of a hazardous substance for use at lower concentrations of that substance, or when required respirator use is independent of concentration.

[2] The assigned protection factors in Table 1 are only effective when the employer implements a continuing, effective respirator program as required by this section (29 CFR 1910.134), including training, fit testing, maintenance, and use requirements.

[3] This APF category includes filtering facepieces and half masks with elastomeric facepieces.

[4] The employer must have evidence provided by the respirator manufacturer that testing of these respirators demonstrates performance at a level of protection of 1,000 or greater to receive an APF of 1,000. This level of performance can best be demonstrated by performing a WPF or SWPF study or equivalent testing. Absent such testing, all other PAPRs and SARs with helmets/hoods are to be treated as loose-fitting facepiece respirators, and receive an APF of 25.

[5] These APFs do not apply to respirators used solely for escape. For escape respirators used in association with specific substances covered by 29 CFR 1910 subpart Z, employers must refer to the appropriate substance-specific standards in that subpart. Escape respirators for other IDLH atmospheres are specified by 29 CFR 1910.134 (d)(2)(ii)

☐ All respirators used shall be NIOSH-approved and used in a manner consistent with NIOSH requirements.

☐ The employer shall identify and evaluate the respiratory hazards for the site. This shall include a reasonable estimate of employee exposure to respiratory hazards as well as the contaminant's chemical state and physical form.

☐ The employer shall provide a selection of respirator models and sizes to ensure proper comfort and fit for each user.

☐ The employer shall not allow entry into areas known to contain IDLH (Immediately Dangerous to Life and Health) atmospheres.

☐ Following are the requirements for respirator use in atmospheres that are not IDLH. The respirator shall be appropriate for the chemical state and physical form of the contaminant:

   o For protection against gases and vapors, the employer shall:

      ◊ Provide an atmosphere-supplying respirator or an air-purifying respirator;

      ◊ Replace cartridges as needed per the environment or contamination exposure requires or:

         o According to published recommendations from respirator manufacturers;

         o According to computer-based service life programs;

- o More frequently based on air sampling results; or,
- o If breakthrough of the contaminant is noticed.
- o For protection against particulates, the employer shall provide:
  - ◊ An atmosphere-supplying respirator; or,
  - ◊ An air-purifying respirator with a NIOSH-certified filters under 30 CFR Part 11 as a HEPA (High Efficiency Particulate Air) or 42 CFR Part 84 for particulates.
  - ◊ The employer may exclude voluntary use of filtering facepiece (dust mask) respirators from the requirements of this section.

MEDICAL EVALUATION

☐ The employer shall provide a medical evaluation (questionnaire) to determine an employee's ability to use a respirator. The questionnaire used is from 29 CFR 1910.134, Appendix C. This medical evaluation and a pulmonary function test shall take place <u>before</u> an employee is fit tested or allowed to use a respirator. Employees who choose to voluntarily wear respirators when not required by exposure to contaminants at or above the PEL (permissible exposure limit) shall also be medically qualified prior to being allowed to wear a respirator. This excludes voluntary use of filtering facepiece (dust mask) respirators.

BP: Find a local occupational clinic that can be used to administer and/or review the questionnaire provided by you, the employer. Administration of the questionnaire consists of providing the questionnaire and an area of privacy to complete the information. If the employer is to administer the questionnaire, make sure the employee's privacy is maintained by providing a private location for completing the questionnaire and a method to seal and deliver the document to the Physician or other Licensed Health Care Professional (PLHCP). We recommend referencing the Health Insurance Portability and Accountability Act (HIPAA) regulations and seeking further advice from a PLHCP.

☐ If a "Yes" response is given to any of questions 1-9, the PLHCP may request a follow-up review of the questions with the employee. This review will determine the necessity for a respiratory physical exam in addition to the pulmonary function test. The employer shall provide the pulmonary function test (PFT) and follow-up respiratory physical exam at no cost if the employee's job duties require the use of a respirator. However, the employer is not obligated to provide follow-up exams for employees who "voluntarily" use respirators. If the employer opts not to provide follow-up exams for such employees, the employee may be given the option of gaining approval by obtaining a follow-up medical evaluation and approval.

☐ The medical questionnaire and examinations shall be administered during normal work hours and in a manner that ensures the employee understands its content.

☐ Employees shall be given the opportunity to discuss the questionnaire and examination results with the PLHCP.

☐ In determining the employee's ability to wear a respirator, the employer shall obtain a written recommendation or disqualification from the PLHCP, if a follow-up medical evaluation is required. This will include:

- o Limitations placed on the employee;
- o The need for follow-up medical evaluations, if required; and
- o A statement that the employee has received a copy of the written recommendation.

☐ Additional medical evaluations shall be conducted for each employee if:

- o The employee reports signs or symptoms related to the use of a respirator;

- o The PLHCP, Respiratory Program Administrator, or the employee's supervisor believes the employee should be re-evaluated; or,

- o Changes in workplace conditions occur that indicate a substantial increase in the physiological burden on the employee.

FIT TESTING

☐ The employer shall fit test all employees with the same make, model, style, and size respirator to be used.

☐ The employer shall utilize a fit test protocol for tight-fitting facepiece respirators that complies with 29 CFR 1910.134, Appendix A.

☐ The employer shall ensure employees using tight-fitting facepiece respirators are fit tested prior to initial use, whenever a different respirator facepiece is used, or at least annually thereafter.
BP: Establish a method to track this annual requirement.

☐ The employer shall perform additional fit tests for employees when the PLHCP, Respiratory Program Administrator, or a supervisor notices a change in the employee's condition that affects respirator fit. Such conditions include:

- o Facial scarring

- o Dental changes

- o Cosmetic surgery; and

- o Obvious change in body weight

☐ The employee shall be given a reasonable opportunity to select another respirator if the respirator being fit is unacceptable.

USE OF RESPIRATORS

☐ Facial hair that may interfere with the seal of the respirator facepiece is prohibited.

☐ If corrective glasses must be worn with the respirator, the corrective lenses may not interfere with the seal of the facepiece.

☐ Employees shall be instructed and required to perform a user seal check each time they put on a tight-fitting facepiece respirator.

☐ The employer shall periodically evaluate effectiveness of respirators.

☐ The employer shall not allow entry to known IDLH atmospheres without the proper training and either supplied-air or SCBA.

BP: Because an annual review of the written respiratory program is required, complete the respirator effectiveness review at this time for periodic use and more frequently for areas that use respirators regularly. Most small employers will not allow employees to perform interior structural fire fighting duties other than incipient-stage fire extinguisher use. Otherwise, further training and information in the written respiratory protection program will be required.

MAINTENANCE, USE, AND INSPECTION OF RESPIRATORS

☐ Utilize and follow the manufacturer's guidelines to ensure all respirators are maintained in a clean and sanitary condition.

☐ All respirators shall be inspected before each use and during cleaning.

☐ Respirators not meeting the requirements of each inspection shall be removed from service and replaced immediately.

BP: As a part of the annual program review, ensure respirators are maintained in a manner consistent with the requirements of the manufacturer.

IDENTIFICATION OF FILTERS AND CARTRIDGES

☐ All filters, cartridges, and canisters are labeled and color-coded with the NIOSH-approval label. This label must not be removed and must remain legible at all times. An example of the use of color-coding labels is a black label band on a cartridge indicating the filters are used for organic vapors. HEPA filters (P100) are identified by a magenta color and protect against particulate dusts from polishing and sanding.

ANNUAL TESTING

☐ Employees who use respirators as a part of the written program must be able to <u>demonstrate</u> understanding of the following:

- o Why the respirator is necessary;

- o How improper fit, usage, or maintenance can compromise the protective effect of the respirator;

- o The limitations and capabilities of the respirator;

- o How to use the respirator in emergency situations, including malfunction of the respirator;

- o How to put on, take off, use, and check the seals of the respirator;

- o Procedures for maintenance and storage of the respirator; and,

- o How to recognize medical symptoms of exposure to the contaminant for which the respirator is being used.

☐ The employer shall ensure this training is provided:

- o Understandably to employees;

- o Prior to the required use of a respirator;

- o Annually for employees using respirators;

- o When changes in the workplace make the previous training obsolete; or,

- o When inadequacies in an employee's knowledge are suspected.

PROGRAM EVALUATION

☐ The Respiratory Protection Program Administrator shall evaluate the program for the following by employee reaction/comment regarding:

   o Respirator fit;

   o Appropriate respirator selection;

   o Respirator use;

   o Respirator effectiveness;

   o Respirator maintenance; and

   o Any additional problems.

☐ Points of comment or concern should be addressed to determine if a revision of the program is necessary.

VOLUNTARY USE OF RESPIRATORS

☐ Employees who choose to wear a respirator (other than filtering facepiece respirators) when not required by exposure to contaminants must be:

   o Medically qualified by the provisions of this program;

   o Trained as required by this program;

   o Properly fit-tested; and

   o Provided a copy of 29 CFR 1910.134 Appendix D upon request.

> **NOTE:** There are no requirements for employees who volunteer to wear only filtering facepiece (dust mask) respirators when contaminant levels are below permissible exposure levels. However, information should always be provided to these employees upon request.

RECORDKEEPING

☐ The employer shall maintain (or coordinate maintenance) the following written records:

   o Medical evaluations (for a period of employment plus 30 years);

   o Training (most current continuously maintained);

   o Fit testing; and

   o Annual program evaluations (maintained for a period of at least 5 years).

------------- END OF SECTION ------------

Section 30
Sanitation

OVERVIEW

This section provides minimum requirements to ensure employees have sufficient clean facilities to perform normal duties, maintain personal hygiene, and consume food. Housekeeping plays a vital role in any effective safety program. A clean and orderly workplace promotes safety, quality, and efficiency on the job.

REGULATORY COMPLIANCE

☐ §1910.141 Sanitation

HOUSEKEEPING

☐ All places of employment must be kept clean, neat, and orderly.

☐ The floor of every workroom must be kept dry when possible.

☐ Drainage systems are required in areas where wet processes are used.

☐ Waterproof footgear, false floors, platforms, mats, or other dry standing places must be provided in wet work areas.

☐ Every floor, workplace, and passageway must be kept free from protruding nails, splinters, loose boards, and unnecessary holes and openings.

WASTE DISPOSAL

☐ Receptacles used for putrid solid or liquid waste refuse must be leak-proof.

☐ Waste receptacles must be maintained in a sanitary condition.

☐ Every waste receptacle must have a solid tightly fitting cover, unless it can be maintained in a sanitary condition without a cover.

☐ Waste receptacles must be constructed of smooth, corrosion-resistant, easily cleanable, or disposable materials.

☐ The number, size, and location of such receptacles must be sufficient to encourage use.

☐ Waste receptacles must not be allowed to overflow.

☐ All sweepings, solid or liquid wastes, refuse, and garbage must be removed as often as necessary and in such a manner as to avoid creating a menace to health.

VERMIN CONTROL

☐ Every enclosed workplace must be kept free of rodents, insects, and other vermin.

☐ A continuing and effective extermination program shall be instituted where the presence of vermin is detected.

WATER SUPPLY

☐ Potable water shall be provided for the following uses:

  o Drinking

  o Washing of persons

  o Cooking

  o Washing of foods

  o Washing of cooking or eating utensils

  o Washing of food preparation or processing premises

  o Personal service rooms

☐ Potable drinking water dispensers shall be designed, constructed, and serviced so that sanitary conditions are maintained.

☐ Potable water shall be equipped with a tap.

☐ Common drinking cups shall be prohibited.

☐ Non-potable water

  o Outlets for non-potable water, such as water for industrial or fire fighting purposes, must be posted or marked in a manner that clearly indicates the water is unsafe to use for consumption.

  o Construction of non-potable water systems shall prevent back flow into potable water systems.

TOILET FACILITIES

Figure 1: Toilet facilities required

Number of Employees	Minimum Number of Water Closets[1]
1 to 15	1
16 to 35	2
36 to 55	3
56 to 80	4
81 to 110	5
111 to 150	6
Over 150	(see note 2)

1. Where women will not use toilet facilities, urinals may be provided instead of water closets, except that the number of water closets in such cases shall not be reduced to less than 2/3 of the minimum specified.
2. One additional fixture for each additional 40 employees.

☐ This section does not apply to mobile crews or normally unattended work locations so long as employees at these locations have immediate transportation to nearby toilet facilities.

☐ Toilet facilities must be provided for each sex unless toilet rooms will be occupied by no more than one person at a time and can be locked from the inside.

☐ Each water closet must be separated with a door and wall or partition to assure privacy.

WASHING FACILITIES

☐ This section does not apply to mobile crews or normally unattended work locations so long as employees at these locations have immediate transportation to nearby toilet facilities.

☐ Washing facilities must be maintained in a sanitary condition.

☐ Lavatories (sinks) must be provided and have hot and cold or tepid running water.

☐ Hand cleaning agents (soap) must be provided.

☐ Paper towels or warm air blowers must be provided.

SHOWERS

☐ If showers are provided, one shower must be maintained for each 10 employees of each sex.

☐ Soap or cleaning agents must be provided.

☐ Showers must have hot and cold water feeding into a common discharge line.

☐ Employees who use showers must be provided with individual clean towels.

CHANGE ROOMS

☐ When employees are exposed to harmful substances that can contaminate clothing, change rooms equipped with storage facilities for street clothes shall be provided.

☐ A separate storage facility for protective clothing must be provided.

☐ Contaminated protective clothing provided by the employer shall be washed and dried before each use to minimize employee exposure.

EATING AND DRINKING AREAS

☐ No employee shall be allowed to consume food or beverages in a toilet room or in any area exposed to a toxic material.

☐ No food or beverages may be stored in a toilet room or in an area exposed to a toxic material.

FOOD HANDLING

☐ All employee food service dispensers must be wholesome and free from spoilage.

☐ Food must be processed, prepared, handled, and stored in a manner as to protect against contamination. (Contact the city or county health department for specific requirements.)

---------- END OF SECTION ----------

Section 31
Scaffolds

OVERVIEW

This section provides key requirements for the erection and safe use of tube and coupler scaffolding as well as mobile ladder stands. This is an overview for general requirements only as there are many types of scaffold systems available. *Always consult the scaffold manufacturer for scaffold erection, use, maintenance, and purpose.*

REGULATORY COMPLIANCE

☐ §1910.28 Safety Requirements for Scaffolding

☐ §1910.29 Manually Propelled Mobile Ladder Stands and Scaffolds (Towers)

GENERAL REQUIREMENTS

☐ Footings or anchorages for scaffolds shall be sturdy enough to carry the maximum intended load without settling.

☐ Scaffolds and their components shall be capable of supporting four times the maximum intended load.

☐ Scaffolds shall not be altered or moved while they are in use.

☐ Any weakened or damaged scaffold shall be repaired immediately and not used until repairs are completed.

☐ Consult a scaffolding manufacturer for the specifications and proper building materials required for erecting a safe scaffold. Do not attempt to erect complicated scaffolding without experienced or professional assistance.

☐ Employees shall not work on scaffolds during storms or high winds.

☐ Tools and other materials shall not be allowed to accumulate in quantities that may cause a hazard.

☐ Wire or fiber rope used for scaffold suspension shall be capable of supporting at least six times the intended load.

☐ Pole scaffolds shall be tied to the building or structure at intervals of no more than 25 feet.

☐ An access ladder or equivalent form of safe access shall be provided.

☐ All scaffolding is required to have guardrails and toeboards as described in the "Guarding Floor and Wall Openings" section.

☐ All platforms shall overlap at least 12 inches and be secured from movement.

☐ Overhead protection shall be provided to all employees working on or near scaffolding.

☐ Runners must be erected along the length of the scaffolding, both inside and outside.

☐ All scaffolding shall be cross-braced.

MANUALLY PROPELLED MOBILE LADDER STANDS AND SCAFFOLDS (TOWERS)

☐ The design shall safely sustain the specified load(s).

☐ Ladder stands shall be designed to support at least four times the desired workload.

☐ Exposed surfaces shall be free from sharp edges, burrs, or other safety hazards.

☐ The maximum work height shall not exceed four times the minimum or smallest base dimensions of any mobile ladder stand or scaffold.

☐ The supporting structure shall be rigidly braced.

☐ All scaffold work levels 10 feet or higher shall have a standard railing system and toeboard.

☐ A climbing ladder or stairway shall be provided for proper access and egress.

☐ Wheels or casters shall be designed to support at least four times the design workload.

☐ Only a scaffold manufacturer or his qualified agent shall be permitted to erect or supervise the erection of scaffolds exceeding 50 feet in height unless the structure is approved in writing by a registered professional engineer or erected in accordance with instructions furnished by the manufacturer.

INSPECTIONS

☐ Scaffolding shall be inspected every day before work begins. If any problems are found with the stability of the structure, repairs shall be made before use.

☐ If cracks or warping are found on planks, they shall be replaced before use. Any sharp edges, burrs, or other safety hazard shall be eliminated.

----------END OF SECTION----------

Section 32
Stairs

OVERVIEW

This section contains specifications for the safe design and construction of fixed general industrial stairs. This includes interior and exterior stairs around machinery, tanks, and other equipment. It also includes stairs leading to and from platforms, floors, and pits.

This section does not apply to stairs used for fire exits, construction, private residences, or to articulated stairs that may be used on floating tanks or on dock facilities.

REGULATORY COMPLIANCE

☐ §1910.24 Fixed Industrial Stairs

WHERE FIXED STAIRS ARE REQUIRED

☐ When routine access between floors is required, fixed industrial stairs must be provided.

☐ Fixed stairs must be provided for access to platforms, tanks, towers, overhead traveling cranes, and other similar structures when possible. (Fixed ladders may be provided if stairs are impractical.)

☐ When employees perform overhead maintenance, inspections, or gauging on machinery, fixed stairs for their path of travel must be provided if fixed ladders are hazardous.

☐ Employees exposed to gases, caustics, acids, or other harmful substances must have fixed stairs for their path of travel.

☐ Spiral stairways are not permitted unless a conventional stairway is impractical.

☐ Winding stairways can be used on structures with a diameter no less than 5 feet (e.g., tanks or similar round structures).

☐ All fixed stairs must have a minimum width of 22 inches.

STAIR STRENGTH

☐ Fixed stairways shall be designed carry a load five times the normal anticipated load while being capable of carrying a moving concentrated load of 1,000 pounds.

ANGLE OF STAIRWAY RISE

☐ Fixed stairs shall have an angle no less than 30 degrees and no more than 50 degrees to the horizontal plane.

Angle to Horizontal	Rise (in inches)	Tread Run (in inches)
30 deg. 35'	6 1/2	11
32 deg. 08'	6 3/4	10 3/4
33 deg. 41'	7	10 1/2
35 deg. 16'	7 1/4	10 1/4
36 deg. 52'	7 1/2	10
38 deg. 29'	7 3/4	9 3/4
40 deg. 08'	8	9 1/2
41 deg. 44'	8 1/4	9 1/4
43 deg. 22'	8 1/2	9
45 deg. 00'	8 3/4	8 3/4
46 deg. 38'	9	8 1/2
48 deg. 16'	9 1/4	8 1/4
49 deg. 54'	9 1/2	8

STAIR TREADS

☐ All stair treads shall have a reasonably slip-resistant surface and nosing and be free of lips or projections. The front portion of steps shall have a non-slip finish.

☐ Grated treads are acceptable as long as personnel can readily identify the leading edge of the tread and the treads have a non-slip finish.

☐ Tread width and rise height shall be uniform throughout all flights of stairs including treads used for part of the foundation.

STAIRWAY PLATFORMS

☐ Platforms shall not be less than the width of the stairway and have a minimum length of 30 inches measured in the direction of travel.

RAILINGS AND HANDRAILS

☐ Any flight of stairs with four or more risers shall have standard railings or standard handrails.

☐ Stairways less than 44 inches wide with both sides enclosed shall have at least one handrail on the right side descending.

☐ Stairways less than 44 inches wide with one side open shall have at least one handrail on the open side.

☐ Stairways less than 44 inches wide with both sides open shall have handrails on both sides.

☐ Stairways more than 44 inches wide but less than 88 inches shall have a handrail on each enclosed side and a handrail on each open side.

☐ Stairways 88 inches or more in width shall have a handrail on each side and one approximately midway of the width.

☐ Winding stairs shall have a handrail that is offset when tread widths are less than 6 inches.

☐ All handrails and toprails shall have a smooth finish throughout the length.

☐ Railings shall have a post, toprail, midrail, and a toeboard. The midrail shall be half way between the toprail and the floor. The toeboard shall be 4 inches high and no more than 1/4 of an inch off the floor.

☐ Top railings shall be no less than 42 inches in height; midrails shall be half the height of the toprail.

☐ To prevent a projection hazard, rails and handrails shall not overhang the end post.

☐ Railings on stairs shall not be less than 30 inches in height or more than 34 inches in height from top of tread to top of handrail.

☐ Wood railing posts shall be constructed of at least 2 x 4-inch stock placed no more than 6 feet apart. Tops and midrails shall also be constructed of 2 x 4-inch stock.

☐ Pipe railings shall be at least 1.5 inches in diameter for posts, toprails, midrails, and toeboards. Posts shall be no more than 8 feet apart.

☐ All posts, toprails, midrails, and toeboards shall be able to withstand a 200-pound load applied in any direction.

☐ Material that is stacked higher than the toeboard shall have paneling from the midrail or to the toprail depending on the height of material.

VERTICAL CLEARANCE

☐ Vertical clearance above any stair tread to an overhead obstruction shall be at least 7 feet measured from the leading edge of the tread.

---------- END OF SECTION ----------

Section 33
Ventilation

OVERVIEW

This section describes the various types of ventilation, ventilation systems, and various operations that require ventilation. It is important to seek the assistance of registered or certified professional industrial hygienists when designing ventilation systems in the workplace.

REGULATORY COMPLIANCE

☐ §1910.94 Ventilation

OPERATIONS THAT MAY REQUIRE VENTILATION

☐ Dust hazards from abrasive blasting

☐ Blast cleaning enclosures

☐ Organic abrasives which are combustible

☐ Areas where particulate fibers are present

☐ Nuisance dust hazards

☐ All areas where fumes are present that may approach a permissible exposure limit (PEL)

☐ Grinding, polishing, and buffing operations

☐ Confined spaces

☐ Areas known to be contaminated with dust or fumes (toxic or not)

☐ Spray areas may require mechanical ventilation adequate to remove flammable vapors, mists, or powders to a safe location and to confine and control combustible residues so that health is not endangered.

VENTILATION INSTALLATION REQUIREMENTS

☐ Ductwork used for ventilation purposes shall not be connected to ducts ventilating any other process.

☐ Where ductwork passes through a combustible roof or wall, the duct shall be made of fire-resistant material.

☐ Inspection or clean-out doors shall be provided for every 9 to 12 feet of running length for ducts up to 12 inches in diameter. The distance may be greater for larger pipes.

☐ A clean-out door or doors shall be provided for servicing the fan and, where necessary, a drain shall be provided.

☐ All ductwork shall be supported to withstand its own weight plus any accumulation.

☐ Testing should be done on the ventilation system before any operation takes place in an area where oxygen concentration is less than 19.5%.

☐ Total air volume exhausted through a spray booth shall dilute the solvent vapor to at least 25% of the Lower Explosive Limit (LEL).

☐ To determine the LEL of the most common solvents used in spray finishing, consult an industrial hygienist, a ventilation engineer or specialist (manufacturer), or use Table G-11 in the Title 29 CFR 1910.94.

☐ It is recommended that an industrial hygienist be consulted to ensure adequate ventilation is provided.

☐ For ventilation in locations containing flammable liquids and vapors, refer to the NFPA Life Safety Code Manual and the Flammable Liquids section.

TYPES OF VENTILATION SYSTEMS

☐ Open air ventilation

☐ Constant air flow systems

☐ Hoods and branch pipes (a system for separating solid contaminates from the air flowing in the system and a discharge stack to the outside)
- o The hood and enclosure design of grinding and cut-off wheel hoods protects the operator from hazards such as bursting wheels. Also, it will provide a means for the removal of dust and dirt generated.
- o Hoods shall be developed to adapt to the particular machine in question.
- o Hoods shall be located as close to the operation as possible.

☐ Exhaust fans with a discharge stack
- o Exhaust systems shall be provided with suitable dust collectors.
- o Exhaust systems shall be tested periodically.

EXHAUST SYSTEMS

☐ Local exhaust systems are usually the proper method of contaminant control if the following situations exist:
- o State or city codes require local exhaust systems for that particular process.
- o Air samples show the contaminant poses a health, fire, or explosion hazard.
- o Maintenance of production machinery would otherwise be difficult.
- o A marked improvement in housekeeping would result.
- o Emission sources are near the employee's breathing zone.
- o Emission sources are large, few, fixed, and/or widely dispersed.
- o Emission rates vary widely by time.

- [ ] Local exhaust systems consist of these major components:
  - o Hood: draws contaminant into the exhaust system.
  - o Duct: carries the contaminant to a central point.
  - o Air-cleaning device: air purifier, filter.
  - o Fan: creates the airflow through the system.
  - o Stack: disperses remaining air contaminants.

- [ ] There are three basic types of hoods: capture, enclosing, and receiving or canopy.

- [ ] A good hood exhaust system includes these features:
  - o An operation enclosed as much as possible to reduce the rate of airflow needed to control the contaminant.
  - o A hood location that pulls the contaminant away from the operator.
  - o A hood located as close as possible to the contaminant.
  - o A hood design that eliminates interference with the worker or the mechanics of the operation.
  - o A hood design that complies with environmental regulations.
  - o Hood location and shape that use the initial velocity of the contaminant to draw it into the hood opening.

- [ ] Choose the correct air-cleaning device for your system. There are several types of these filters. Each is designed for filtering a specific contaminant (e.g., air cleaners for fumes or smokes, air cleaners for radioactive dusts, or air cleaners for gases and vapors). Consideration should be given to two factors when installing an air cleaner: its efficiency and performance rating. These factors also include the following considerations:
  - o Rate of resistance build-up.
  - o Dust-holding capacity.
  - o Ability to be easily cleaned (in permanent filters).
  - o Longevity (in disposable filters).

- [ ] Fans shall be grounded in areas where flammable dusts or fumes are being ventilated. The fan shall be approved for the particular conditions or hazard.

---------- END OF SECTION ----------

Section 34
Welding, Cutting, and Brazing

OVERVIEW

This section describes key safety precautions necessary for various methods of welding, cutting, and brazing.

REGULATORY COMPLIANCE

☐ 29 Code of Federal Regulations (CFR), Subpart Q – Welding, Cutting, and Brazing, §1910.251-255

PRECAUTIONS FOR FIRE PREVENTION

☐ If the object being welded or cut cannot be moved, all moveable fire hazards shall be moved to a safe place.

☐ If the object being welded or cut cannot be moved, and if all fire hazards cannot be removed, then guards shall be used to protect the immovable fire hazards.

☐ Special precautions shall be taken to remove or cover combustible materials that may be exposed to sparks wherever there are holes or cracks in floors, walls, windows, etc.

☐ Suitable fire extinguishing equipment (portable fire extinguishers, pails of water, buckets of sand, etc.) shall be readily available for instant use.

☐ Require firewatchers whenever welding or cutting is performed in locations other than where minor fire might break out.

☐ Firewatchers shall have fire-extinguishing equipment readily available and must be trained in its use.

☐ Whenever firewatchers are required, the fire watch shall be maintained for at least a half hour after completion of welding and cutting operations to detect and extinguish possible smoldering fires.

☐ Before welding or cutting is permitted, a person responsible for authorizing welding and cutting operations shall inspect the area and designate precautions to be followed, preferably in the form of a written permit.

☐ Floors shall be kept clear of combustible materials (paper, wood chips, etc.).

☐ Combustible floors shall be kept wet, covered by damp sand, or covered by fire-resistant shields. Where floors have been wetted, personnel operating arc welding or cutting equipment shall be protected from shock.

- [ ] Welding and cutting shall be prohibited in these areas:

  - Areas not authorized by management

  - Buildings with impaired sprinkler systems

  - Areas with explosive or potentially explosive atmospheres

  - Storage areas with large quantities of exposed, readily ignitable materials such as bulk sulfur, baled paper, or cotton

- [ ] Relocate combustibles at least 35 feet from the work site or protect with flameproof covers or shields.

- [ ] Ducts that might carry sparks to combustibles shall be suitably covered or shut down.

- [ ] Management shall recognize its responsibility for the safe use of cutting and welding equipment on its property. Management responsibilities include:

  - Establishing procedures and areas for welding and cutting operations

  - Designating an individual responsible for authorizing welding and cutting operations in areas not specifically designed for such processes

  - Ensuring that cutters or welders and supervisors are suitably trained and judged competent

  - Warning contractors of combustible hazards that may not be obvious

- [ ] The supervisor shall be responsible for the following:

  - The safe use of the cutting or welding process

  - Determining the combustible or hazardous materials present or likely to be present

  - Ensuring protection of combustibles from ignition

  - Securing authorization for the welding or cutting operation

  - Ensuring that the welder or cutter secures assurance of safe conditions

  - Determining that fire protection and extinguishing equipment is properly located at the site

  - Ensuring that fire watchers are at the site when required

- [ ] No welding, cutting, or other hot work shall be performed on used drums, barrels, tanks, or other containers until they have been cleaned so thoroughly that there are no flammable materials present.

PROTECTION OF PERSONNEL

- [ ] Operators shall use any necessary personal protective equipment to prevent injury. Such equipment may include, but is not limited to the following:

  - Fall protection equipment, where applicable

  - Helmets

  - Hand shields

  - Goggles (ventilated) or other suitable eye protection

  - Filter lenses

  - Tempered lenses

- o   Respirator equipment
- o   Protective clothing
- o   Ventilation
- o   Lifelines for confined spaces

OXYGEN-FUEL GAS WELDING

☐ Only approved apparatus shall be used (torches, regulators or pressure-reducing valves, acetylene generators, manifolds).

☐ Cylinders shall be kept away from radiators and other sources of heat, in a well-protected, ventilated, dry location at least 20 feet from highly combustible materials such as oil. They shall be stored in definite assigned places away from elevators, stairs, or gangways and where they will not be knocked over or damaged by passing or falling objects or tampered with by unauthorized persons.

☐ Empty cylinders shall have their valves closed and valve protection caps installed for storage.

☐ Oxygen cylinders or apparatus shall not be handled with oily hands or gloves.

ARC WELDING AND CUTTING

☐ For arc welding under wet conditions (including heavy perspiration), the use of reliable automatic controls for reducing no-load voltage is recommended to reduce the shock hazard.

☐ Control apparatus shall be enclosed except for the operating wheels, levers, or handles.

☐ The frame or case of the welding machine (except engine-driven machines) shall be grounded.

☐ Before beginning a welding operation, perform the following tasks to ensure safe operation of the equipment:
- o   Spread out coiled welding cable before use to avoid serious overheating and damage to insulation.
- o   Check the grounding of the welding machine frame.
- o   Check for leaks of cooling water, shielding gas, or engine fuel.
- o   Ensure that proper switching equipment for shutting down the machine is provided.
- o   Ensure that manufacturers' instructions are followed.
- o   Ensure that electrode holders cannot make electrical contact with personnel, conducting objects, fuel, or compressed gas tanks when not in use.
- o   Ensure that there are no splices within 10 feet of the holder.
- o   The welder should not coil or loop electrode cable around parts of his body.
- o   The operator shall report any equipment defect or safety hazard to his supervisor and discontinue use of the equipment until repaired.

----------END OF SECTION----------

Appendix A
OSHA Inspection

The Williams-Steiger Occupational Safety and Health Act of 1970 requires, in part, that every employer covered under the Act furnish to his/her employees employment and a place of employment which is free from recognized hazards that are causing or are likely to cause death or serious physical harm to his/her employees.

The Act also requires that employers comply with occupational safety and health standards promulgated under the Act, and that employees comply with standards, rules, regulations, and orders issued under the Act which are applicable to their own actions and conduct.

The Act authorized the Department of Labor to conduct inspections, and to issue citations and proposed penalties for alleged violations.

The Act, under section 20 [B], also authorizes the Secretary of Health, Education, and Welfare to conduct inspections and to question employers and employees in connection with research and other related activities.

The Act contains provisions for adjudication of violations and proposed penalties by the occupational safety and health review commission, if contested by an employer or by an employee or authorized representative of employees. The Act also contains provisions for judicial review.

The purpose of part 1903 is to prescribe rules and to set forth general policies for enforcement of the inspection, citation, and proposed penalty provisions of the Act.

In situations where part 1903 sets forth general enforcement policies rather than substantive or procedural rules, such policies may be modified in specific circumstances where the secretary or his designee determines that an alternative course of action would better serve the objectives of the Act.

Authority for Inspection

Compliance safety and health officers of the Department of Labor are authorized to enter without delay and at reasonable times any factory, plant, establishment, construction site, or other area workplace or environment where work is performed by an employee of any employer; to inspect and investigate during regular working hours and at other reasonable times, within reasonable limits and in a reasonable manner, any such place of employment, and all pertinent conditions, structures, machines, apparatus, devices, equipment, and materials therein; to question privately any employer, owner, operator, agent, or employee; and to review records required by the Act and regulations published, or other records which are directly related to the purpose of the inspection. Representatives of the Secretary of Health, Education, and Welfare are authorized to make inspections and to question employers and employees in order to carry out the functions of the Secretary of Health, Education, and Welfare under the Act.

Inspections conducted by Department of Labor compliance safety and health officers and representatives of the Secretary of Health, Education, and Welfare under section 8 of the Act and pursuant of this part 1903 shall not affect the authority of any state to conduct inspections in accordance with agreements and plans under section 18 of the Act.

Mechanics of OSHA Inspection

1. **Entry of the Workplace**
   a) Advance Notice
   b) Time of Inspection
   c) Present
   d) State Why and What Kind of Inspection

2. **Opening Conference**
   a) Keep Conference Brief
   b) Conducted in the Open at Worksite
   c) Outline Scope of Inspection
      1) Records - OSHA 300, 300-A & 301
      2) Employee Interviews
      3) Safety Committee Meetings Minutes
      4) Trade Secrets
      5) Photographs and Tape Recorder
      6) Employees of Other Employers

3. **Walk-Around Representatives**
   a) Employer Rep [Safety Director]
   b) Employee Rep
      1) Union
      2) Safety Committee
   c) Non Authorized Rep—Random Selection

4. **Walk-Around Inspection**
   a) Standards Compliance
   b) General Duty Clause

5. **Closing Conference**
   a) Advise of Apparent Violations
   b) Discuss Citation and Penalties
   c) Abatement

Appendix B
OSHA Regional Offices

In case of emergency, call 1-800-321-OSHA (6742)

Region 1
CT, MA, ME, NH, RI, VT
JFK Federal Building, Room E340
Boston, MA 02203
Telephone: (617) 565-9860
Fax: (617) 565-9827

Region 2
NJ, NY, Puerto Rico, Virgin Islands
201 Varick Street, Room 670
New York, NY 10014
Telephone: (212) 337-2378
Fax: (212) 337-2371

Region 3
DC, DE, MD, PA, VA, WV
The Curtis Center – Suite 740 West
170 S. Independence Mall West
Philadelphia, PA 19106-3309
Telephone: (215) 861-4900
Fax: (215) 861-4904

Region 4
AL, FL, GA, KY, MS, NC, SC, TN
61 Forsyth Street, SW
Atlanta, GA 30303
Telephone: (404) 562-2300
Fax: (404) 562-2295

Region 5
IL, IN, MI, MN, OH, WI
230 South Dearborn Street, Room 3244
Chicago, IL 60604
Telephone: (312) 353-2220
Fax: (312) 353-7774

Region 6
AR, LA, NM, OK, TX
525 Griffin Street, Room 602
Dallas, Texas 75202
Telephone: (972) 850-4145
Fax: (972) 850-4149

Region 7
IA, KS, MO, NE
City Center Square
1100 Main Street, Suite 800
Kansas City, MO 64105
Telephone: (816) 426-5861
Fax: (816) 426-2750

Region 8
CO, MT, ND, SD, UT, WY
1999 Broadway, Suite 1690
Denver, CO 80201-6550
Telephone: (720) 264-6550
Fax: (720) 264-6585

Region 9
AZ, CA, Guam, HI, NV
71 Stevenson Street, Room 420
San Francisco, CA 94105
Telephone: (415) 975-4310
Fax: (415) 975-4319
NOTE: For federal agencies or private
companies working for federal agencies
only. AZ, CA, HI, and NV have State offices.

Region 10
AK, ID, OR, WA
1111 Third Avenue, Suite 715
Seattle, WA 98101-3212
Telephone: (206) 553-5930
Fax: (206) 553-6499

Appendix C
State Plan Programs

What is a State OSHA Program?

Section 18 of the Occupational Safety and Health Act of 1970 (the Act) encourages States to develop and operate their own job safety and health programs. OSHA approves and monitors State plans and provides up to 50 percent of an approved plan's operating costs.

There are currently 22 States and jurisdictions operating complete State plans (covering both the private sector and State and local government employees) and four - Connecticut, New Jersey, New York, and the Virgin Islands - which cover public employees only. (Eight other States were approved at one time but subsequently withdrew their programs.)

(Please note that the Connecticut, New Jersey, New York, and Virgin Islands plans cover public sector employment only.)

States must set job safety and health standards that are "at least as effective as" comparable federal standards. (Most States adopt standards identical to federal ones.) States have the option to promulgate standards covering hazards not addressed by federal standards.

A State must conduct inspections to enforce its standards, cover public (State and local government) employees, and operate occupational safety and health training and education programs. In addition, most States provide free on-site consultation to help employers identify and correct workplace hazards. Such consultation may be provided either under the plan or through a special agreement under section 21(d) of the Act.

How does a State establish its own program?

To gain OSHA approval for a **developmental plan** - the first step in the State plan process - a State must assure OSHA that within 3 years it will have in place all the structural elements necessary for an effective occupational safety and health program. These elements include: appropriate legislation; regulations and procedures for standards setting, enforcement, appeal of citations and penalties; a sufficient number of qualified enforcement personnel.

Once a State has completed and documented all its developmental steps, it is eligible for **certification**. Certification renders no judgment as to actual State performance, but merely attests to the structural completeness of the plan.

At any time after initial plan approval, when it appears that the State is capable of independently enforcing standards, OSHA may enter into an **"operational status agreement"** with the State. This commits OSHA to suspend the exercise of discretionary federal enforcement in all or certain activities covered by the State plan.

The ultimate accreditation of a State's plan is called **final approval**. When OSHA grants final approval to a State under Section 18 (e) of the Act, it relinquishes its authority to cover occupational safety and health matters covered by the State. After at least 1 year following certification, the State becomes eligible for final approval if OSHA determines that it is providing, in actual operation, worker protection "at least as effective" as the protection provided by the federal program. The State also must meet 100 percent of the established compliance staffing levels (benchmarks) and participate in OSHA's computerized inspection data system before OSHA can grant final approval.

Employees finding workplace safety and health hazards may file a formal complaint with the appropriate plan State or with the appropriate OSHA regional administrator. Complaints will be investigated and should include the name of the workplace, type(s) of hazard(s) observed, and any other pertinent information.

Anyone finding inadequacies or other problems in the administration of a State's program, may file a Complaint About State Program Administration (CASPA) with the appropriate OSHA regional administrator as well. The complainant's name is kept confidential. OSHA investigates all such complaints, and where complaints are found to be valid, requires appropriate corrective action on the part of the State.

Directory of States with Approved Occupational Safety and Health Plans

Alaska Department of Labor and Workforce Development
P.O. Box 111149
Juneau, Alaska 99811-1149
1111 W. 8th Street, Room 308
Juneau, Alaska 99801-1149
Greg O'Claray, Commissioner (907) 465-2700 Fax: (907) 465-2784
Grey Mitchell, Director (907) 465-4855 Fax: (907) 465-6012

Industrial Commission of Arizona
800 W. Washington
Phoenix, Arizona 85007-2922
Larry Etchechury, Director, ICA (602) 542-4411 Fax: (602) 542-1614
Darin Perkins, Program Director (602) 542-5795 Fax: (602) 542-1614

California Department of Industrial Relations
1515 Clay Street, Suite 1901
Oakland, California 94612
John Rea, Acting Director (415) 703-5050 Fax:(415) 703-5059
Len Welsh, Acting Chief, Cal/OSHA (510) 286-7000 FAX (510) 286-7038
Vicky Heza, Deputy Chief, Cal/OSHA (714) 939-8093 FAX (714) 939-8094

Connecticut Department of Labor
200 Folly Brook Boulevard
Wethersfield, Connecticut 06109
Patricia H. Mayfield, Commissioner (860) 566-5123 Fax: (860) 566-1520
Conn-OSHA
38 Wolcott Hill Road
Wethersfield, Connecticut 06109
Richard Palo, Director (860) 263-6900 Fax: (860) 263-6940

Hawaii Department of Labor and Industrial Relations
830 Punchbowl Street
Honolulu, Hawaii 96813
Nelson B. Befitel, Director (808) 586-8844 Fax: (808) 586-9099

Indiana Department of Labor
State Office Building
402 West Washington Street, Room W195
Indianapolis, Indiana 46204-2751
Miguel Rivera, Commissioner (317) 232-2378 Fax: (317) 233-3790
Tim Grogg, Deputy Commissioner (317) 233-3605 Fax: (317) 233-3790

Iowa Division of Labor
1000 E. Grand Avenue
Des Moines, Iowa 50319-0209
Dave Neil, Commissioner (515) 281-3447 Fax: (515) 281-4698
Mary L. Bryant, Administrator (515) 281-3469 Fax: (515) 281-7995

Kentucky Department of Labor
1047 U.S. Highway 127 South, Suite 4
Frankfort, Kentucky 40601
Philip P. Anderson, Commissioner (502) 564-3070 Fax: (502) 564-5387
Stephen L. Morrison, Executive Director, Office of Occupational Safety & Health (502) 564-3070
Fax: (502) 564-1682

Maryland Division of Labor and Industry
Department of Labor, Licensing and Regulation
1100 North Eutaw Street, Room 613
Baltimore, Maryland 21201-2206
Robert Lawson, Commissioner (410) 767-2241 Fax: (410) 767-2986
Jack English, Assistant Commissioner, MOSH (410) 767-2190 Fax: (410) 333-7747

Michigan Department of Labor and Economic Growth
Robert Swanson, Director
Michigan Occupational Safety and Health Administration
P.O. Box 30643
Lansing, MI 48909-8143
Doug Kalinowski, Director (517) 322-1814 Fax: (517) 322-1775
Martha Yoder, Deputy Director for Enforcement
(517) 322-1817 Fax: (517) 322-1775

Minnesota Department of Labor and Industry
443 Lafayette Road
St. Paul, Minnesota 55155
Scott Brener, Commissioner (651) 284-5010 Fax: (651) 282-5405
Patricia Todd, Assistant Commissioner (651) 284-5371 Fax: (651) 282-2527
Jeff Isakson, Administrative Director, OSHA Management Team
(651) 284-5372 Fax: (651) 297-2527

Nevada Division of Industrial Relations
400 West King Street, Suite 400
Carson City, Nevada 89703
Roger Bremmer, Administrator (775) 684-7260 Fax: (775) 687-6305
Occupational Safety and Health Enforcement Section (OSHES)
1301 N. Green Valley Parkway
Henderson, Nevada 89014
Tom Czehowski, Chief Administrative Officer (702) 486-9168 Fax: (702) 486-9020
[Las Vegas (702) 687-5240]

New Jersey Department of Labor and Workforce Development
Office of Public Employees Occupational Safety & Health (PEOSH)
1 John Fitch Plaza
P.O. Box 386
Trenton, NJ 08625-0386
Thomas D. Carver, Acting Commissioner (609) 292-2975 Fax: (609) 633-9271
Leonard Katz, Assistant Commissioner (609) 292-2313 Fax: (609) 695-1314
Howard Black, Director, PSOSH (609) 292-0501 Fax: (609) 292-3749
Gary Ludwig, Director, Occupational Health Service (609) 984-1843 Fax: (609) 984-0849

New Mexico Environment Department
1190 St. Francis Drive, Suite 4050
P.O. Box 26110
Santa Fe, New Mexico 87502
Ron Curry, Jr., Secretary (505) 827-2850 Fax: (505) 827-2836
Butch Tongate, Bureau Chief (505) 476-8700 Fax: (505) 476-8734

New York Department of Labor
New York Public Employee Safety and Health Program
State Office Campus Building 12, Room 158
Albany, New York 12240
Linda Angello, Commissioner (518) 457-2741 Fax: (518) 457-6908
Anthony Germano, Director, Division of Safety and Health
(518) 457-3518 Fax: (518) 457-1519
Maureen Cox, Program Manager (518) 457-1263 Fax: (518) 457-5545

North Carolina Department of Labor
4 West Edenton Street
Raleigh, North Carolina 27601-1092
Cherie Berry, Commissioner (919) 733-0359 Fax: (919) 733-1092
Allen McNeely, Deputy Commissioner, OSH Director (919) 807-2861 Fax: (919) 807-2855
Kevin Beauregard, OSH Assistant Director (919) 807-2863 Fax: (919) 807-2856

Oregon Occupational Safety and Health Division
Department of Consumer and Business Services
350 Winter Street, NE, Room 430
Salem, Oregon 97301-3882
Michael Wood, Administrator (503) 378-3272 Fax: (503) 947-7461
Michele Patterson, Deputy Administrator (503) 378-3272 Fax: (503) 947-7461
David Sparks, Special Assistant for Federal & External Affairs (503) 378-3272
Fax: (503) 947-7461

Puerto Rico Department of Labor
Prudencio Rivera Martínez Building
505 Muñoz Rivera Avenue
Hato Rey, Puerto Rico 00918
Roman M. Velasco Gonzalez, Secretary
(787) 754-2119 Fax: (787) 753-9550
José Droz-Alvarado, Assistant Secretary for Occupational Safety and Health
(787) 756-1100 / (787) 754-2171 Fax: (787) 767-6051

South Carolina Department of Labor, Licensing, and Regulation
Koger Office Park, Kingstree Building
110 Centerview Drive
PO Box 11329
Columbia, South Carolina 29211
Adrienne R. Youmans, Director (803) 896-4300 Fax: (803) 896-4393
Dottie Ison, Administrator (803) 896-7665 Fax: (803) 896-7670
Office of Voluntary Programs (803) 896-7744 Fax: (803) 896-7750

Tennessee Department of Labor and Workforce Development
710 James Robertson Parkway
Nashville, Tennessee 37243-0659
James G. Neeley, Commissioner (615) 741-2582 Fax: (615) 741-5078
John Winkler, Program Director (615) 741-2793 Fax: (615) 741-3325

Utah Labor Commission
160 East 300 South, 3rd Floor
PO Box 146650
Salt Lake City, Utah 84114-6650
Sherrie M. Hayashi, Commissioner (801) 530-6848 Fax: (801) 530-7906
Larry Patrick, Administrator (801) 530-6898 Fax: (801) 530-6390

Vermont Department of Labor
National Life Building - Drawer 20
Montpelier, Vermont 05620-3401
Patricia Moulton Powden, Commissioner (802) 828-4301 Fax: (802) 888-4022
Vermont OSHA
National Life Building - Drawer 20
Montpelier, Vermont 05620-3401
Robert McLeod, VOSHA Compliance Program Manager (802) 828-2765 Fax: (802) 828-2195

Virgin Islands Department of Labor
3012 Golden Rock
Christiansted, St. Croix, Virgin Islands 00820-4660
Cecil R. Benjamin, Commissioner (340) 773-1994 Fax: (340) 773-1858
John Sheen, Assistant Commissioner (340) 772-1315 Fax: (340) 772-4323
Francine Lang, Program Director (340) 772-1315 Fax: (340) 772-4323

Virginia Department of Labor and Industry
Powers-Taylor Building
13 South 13th Street
Richmond, Virginia 23219
C. Raymond Davenport, Commissioner (804) 786-2377 Fax: (804) 371-6524
Glenn Cox, Director, Safety Compliance, VOSHA (804) 786-2391 Fax: (804) 371-6524
Jay Withrow, Director, Office of Legal Support (804) 786-9873 Fax: (804) 786-8418

Washington Department of Labor and Industries
General Administration Building
PO Box 44001
Olympia, Washington 98504-4001
7273 Linderson Way SW
Tumwater, WA 98501-5414
Gary K. Weeks, Director (360) 902-4200 Fax: (360) 902-4202
Steve Cant, Assistant Director [PO Box 44600]
(360) 902-5495 Fax: (360) 902-5529
Program Manager, Federal-State Operations [PO Box 44600]
(360) 902-5430 Fax: (360) 902-5529

Wyoming Department of Employment
Workers' Safety and Compensation Division
Cheyenne Business Center
1510 East Pershing Boulevard
Cheyenne, Wyoming 82002
Gary W. Child, Administrator (307) 777-7700 Fax: (307) 777-5524
J.D. Danni, OSHA Program Manager (307) 777-7786 Fax: (307) 777-3646

Appendix D
References

U. S. Government Publications

Title 29 CFR Parts 1900.1 to 1910.999

Title 29 CFR Parts 1910.1000 to END

Title 40 CFR Parts 260 to 370 (Hazardous Waste Regulations)

Title 49 CFR Parts 100 to 185 (Hazardous Materials Regulations)

National Safety Council

Accident Prevention Manual for Business and Industry: Administration and Programs, 12th Edition

National Fire Protection Association

NFPA 10: Standard for Portable Fire Extinguishers 2002 Edition

NFPA 30: Flammable and Combustible Liquids Code 2003 Edition

NFPA 51B: Standard for Fire Prevention during Welding, Cutting, and Other Hot Work 2003 Edition

NFPA 55: Standard for the Storage, Use, and Handling of Compressed Gases and Cryogenic Fluids in Portable and Stationary Containers, Cylinders, and Tanks 2005 Edition

NFPA 70: National Electric Code (NEC) 2005 Edition

NFPA 101: Life Safety Code 2006 Edition

NFPA 251: Standard Methods of Tests of Fire Resistance of Building Construction and Materials 2006 Edition

Appendix E
Essential Internet Resources

OSHA Forms and Posters

- Forms for Occupational Injury and Illness Recordkeeping
  www.osha-slc.gov/recordkeeping/RKforms.html

- Form 174 Material Safety Data Sheet (MSDS)
  www.osha-slc.gov/dsg/hazcom/msdsformat.html

OSHA Posters

- Job Safety and Health Protection Poster (OSHA 3165)
  www.osha-slc.gov/Publications/poster.html

- Access to Medical and Exposure Records
  http://www.osha-slc.gov/Publications/pub3110text.html

All of these and more can be found on the OSHA forms and posters index at:

www.osha-slc.gov/pls/publications/pubindex.list#posters1

Online Resources

- OSHA e-Tools web site with several MS PowerPoint Presentations
  www.osha-slc.gov/dts/osta/oshasoft/index.html

- ANSI Online Electronic Standards (fee-based access to ANSI standards)
  www.webstore.ansi.org

- MSDS Collection/Vermont SIRI (free collection of over 100,000 manufacturer-supplied MSDSs)
  www.siri.org/msds

- NIOSH Certified Equipment (directory of NIOSH-approved respiratory protective equipment)
  www.cdc.gov/niosh/94-104.html

- Safety Training Modules (provided by Oklahoma State University)
  http://www.pp.okstate.edu/ehs/modules/index.htm

- OSHA Job Hazard Analysis Guidance

  www.osha-slc.gov/Publications/osha3071.html

Agencies and Organizations

American Conference of Government
Industrial Hygienists, Inc. (ACGIH)
6500 Glenway Ave., Bldg D-7
Cincinnati, OH 45211
(513) 661-7881
www.acgih.org

American Industrial Hygiene
Association (AIHA)
2700 Prosperity Ave., Suite 250
Fairfax, VA 22031
(703) 849-8888
www.aiha.org

American National Standards
Institute (ANSI)
11 West 42nd Street
New York, NY 10036
(212) 642-4900 or (212) 764-3274
www.ansi.org

American Society of Safety
Engineers (ASSE)
1800 East Oakton Street
Des Plaines, IL 60018
(708) 692-4121
www.asse.org

Compressed Gas Association
1725 Jefferson Davis Hwy. Suite 1004
Arlington, VA 22202-4102
(703) 412-0900
www.cganet.com

Government Printing Office (GPO)
Washington, DC 20402-9371
(202) 512-2457
www.gpo.gov

National Fire Protection Assoc. (NFPA)
1 Batterymarch Park
Quincy, MA 02269
(617) 770-3000
www.nfpa.org

National Institute for Occupational
Safety & Health (NIOSH)
4676 Columbia Parkway
Cincinnati, OH 45226
(513) 533-8236
www.cdc.gov/niosh/homepage.html

National Safety Council (NSC)
1121 Spring Lake Drive
Itasca, IL 60143-3201
(708) 285-1121
www.nsc.com

Nuclear Regulatory Commission
One White Flint North Building
11555 Rockville Pike
Rockville, MD 20852
(202) 634-3273
www.nrc.gov

Occupational Safety & Health Administration (OSHA)
200 Constitution Ave, NW
Washington, DC 20210
(202) 634-7960
www.osha.gov

Safety Equipment Institute
1901 N. Moore St., Suite 808
Arlington, VA 22209
(703) 525-1695
www.seinet.org

U.S. Department of
Transportation (DOT)
400 7th St. SW
Washington, DC 20590
(202) 366-4000
www.dot.gov

U.S. Environmental Protection
Agency (EPA)
401 M Street SW
Washington, DC 20460
(202) 260-2090
www.epa.gov

U.S. Food & Drug Administration
5600 Fishers Lane
Rockville, MD 20857
(301) 443-1544
www.fda.gov

Appendix F
Self-Inspection Checklist

GENERAL

☐Yes ☐No   Is the safety committee or group meeting regularly and reporting its activities in writing?

☐Yes ☐No   Are emergency evacuation routes identified?

☐Yes ☐No   Is one person clearly in charge of safety and health activities?

☐Yes ☐No   Is the OSHA poster prominently displayed in your business where all employees are likely to see it?

☐Yes ☐No   Are emergency telephone numbers posted where they can be found readily?

☐Yes ☐No   Is a substance abuse policy in place?

☐Yes ☐No   Is management demonstrating an active interest in safety and health matters by defining a policy for your business and communicating it to all employees?

☐Yes ☐No   Is there a safety committee or group that allows employee participation in safety and health activities?

☐Yes ☐No   Are combustible waste and debris removed from work areas promptly and stored safely?

☐Yes ☐No   Are adequate toilets and washing facilities provided?

OSHA RECORDKEEPING

☐Yes ☐No   Are work-related injuries and illnesses being recorded on the OSHA 300 log?

☐Yes ☐No   Is an annual summary or OSHA 300-A posted from February 1 to April 30 of each year?

☐Yes ☐No   Is OSHA contacted within 8 hours of a fatality or an accident resulting in the hospitalization of five or more employees?

☐Yes ☐No   Are required medical records and exposure records maintained?

☐Yes ☐No   Are required training records maintained?

☐Yes ☐No   Are employee records being maintained for the appropriate length of time?

☐Yes ☐No   Are operating permits and records up to date?

☐Yes ☐No   Are procedures in place to maintain records and logs of safety inspections, safety meeting minutes, accident investigations, and emergency response drills?

TRAINING

☐Yes ☐No   Is safety and health training provided for all employees requiring such training?

☐Yes ☐No   Are adequate training resources available?

☐Yes ☐No   Is management committed to employee training?

☐Yes ☐No   Is all required training adequately documented?

☐Yes ☐No   Do employees participate in regularly scheduled safety meetings?

☐Yes ☐No   Have all employees received training in work area hazards, emergency action plans, equipment operation, personal protective equipment, location and use of emergency equipment, hazard communication (MSDS), and hearing conservation?

☐Yes ☐No   Do all employees receive refresher training at least once per year?

☐Yes ☐No   Have all employees been trained in proper procedures for reporting unsafe conditions, defective equipment, and unsafe acts?

MEDICAL SERVICES/FIRST-AID

☐Yes ☐No Is a medically approved first-aid kit available and stocked adequately?
☐Yes ☐No Is a bloodborne pathogen plan in place?
☐Yes ☐No Is regulated waste discarded according to applicable laws and regulations?
☐Yes ☐No Is a medically approved sharps container adequately supplied?
☐Yes ☐No If medical or first-aid facilities are not available on site, is at least one employee on each shift qualified to administer first aid?
☐Yes ☐No Are qualified medical personnel readily available for advice and consultation?
☐Yes ☐No Are quick drenching and/or flushing areas available in areas where corrosive liquids or materials are handled?

FIRE PROTECTION

☐Yes ☐No Are fire extinguishers inspected monthly for general condition and operability and noted on the inspection tag?
☐Yes ☐No Are fire extinguishers recharged regularly and properly noted on the inspection tag?
☐Yes ☐No Are fire extinguishers mounted in readily accessible locations?
☐Yes ☐No Are portable fire extinguishers provided in adequate number and type?
☐Yes ☐No Are fire alarm systems tested at least annually?
☐Yes ☐No Are plant employees periodically instructed in the use of fire protection procedures?
☐Yes ☐No Are fire hydrants flushed at least once a year and maintained regularly?
☐Yes ☐No Are fire doors and shutters in good operating condition?
☐Yes ☐No Are fire doors and shutters unobstructed and protected against obstruction?
☐Yes ☐No Is your local fire department well acquainted with your plant location and specific hazards?
☐Yes ☐No Are water control valves, air and water pressure checked weekly on automatic sprinklers?
☐Yes ☐No Are control valves locked open on automatic sprinklers?
☐Yes ☐No Is maintenance of the system assigned to a responsible person or a sprinkler contractor?
☐Yes ☐No Are sprinkler heads protected by metal guards where they are exposed to mechanical damage?
☐Yes ☐No Is proper minimum clearance maintained around sprinkler heads?
☐Yes ☐No Are interior standpipes and valves inspected on a regular basis?

WORK AREAS

☐Yes ☐No Are work areas clean and orderly?
☐Yes ☐No Are workers aware of the hazards involved with the various chemicals they may be exposed to?
☐Yes ☐No Is employee exposure to chemicals kept within acceptable levels?
☐Yes ☐No Are proper precautions being taken for handling asbestos and other fibrous materials?
☐Yes ☐No Are appropriate caution labels and signs used to warn employees of hazardous substances?
☐Yes ☐No Is potable water provided for drinking, washing, and cooking?
☐Yes ☐No Are water outlets not suitable for drinking clearly identified?
☐Yes ☐No Are wet methods used to prevent the emission of hazardous airborne fibers?

Ventilation

☐Yes ☐No   Is the building properly ventilated?
☐Yes ☐No   Are all ventilation systems in good working order?
☐Yes ☐No   Are exhaust fans used during welding procedures?
☐Yes ☐No   Are machines that produce dust vented to an industrial collector or central exhaust system?
☐Yes ☐No   Are ventilation systems appropriate for the work being performed?
☐Yes ☐No   Are the volume and velocity of each exhaust system sufficient to gather the fumes, mists, vapors, gases, or dusts to be controlled?
☐Yes ☐No   Are exhausts conveyed to a suitable point of disposal?
☐Yes ☐No   Are clean-out ports or doors provided at intervals of no more than 12 feet in all horizontal runs of exhaust ducts?
☐Yes ☐No   Where two or more different operations are being controlled by the same exhaust system will the combination of substances present the risk of fire, explosion, or chemical reaction in the duct?
☐Yes ☐No   Where two or more ventilation systems are serving the same work area, is their operation such that one will not offset the function or effectiveness of the other?
☐Yes ☐No   Is adequate makeup air provided to areas where exhaust systems are operating?
☐Yes ☐No   Is the source point for makeup air located so that only clean, fresh air, which is free of contaminates, will enter the system?
☐Yes ☐No   Are spray painting operations done in spray rooms or booths equipped with an appropriate exhaust system?
☐Yes ☐No   Are ventilation filters replaced regularly?
☐Yes ☐No   Is the building climate kept at a comfortable level?
☐Yes ☐No   Are confined spaces properly ventilated?
☐Yes ☐No   Are employees adequately protected from heat stress and/or hypothermia?

Walkways

☐Yes ☐No   Are work areas free of slip and fall hazards?
☐Yes ☐No   Are the floors in work areas dry and free of oil?
☐Yes ☐No   Are aisles properly marked?
☐Yes ☐No   Are slip resistant mats used in wet areas?
☐Yes ☐No   Are uneven floor elevations marked?
☐Yes ☐No   Are floor openings guarded and marked?
☐Yes ☐No   Are warning signs posted and clearly visible?
☐Yes ☐No   Are pedestrians and other traffic protected from open trenches?
☐Yes ☐No   Are employees protected from high traffic areas?
☐Yes ☐No   Are walkways arranged to minimize hazard exposure from operating machinery, welding operations, or similar operations?
☐Yes ☐No   Are fork truck lanes properly marked?
☐Yes ☐No   Is adequate headroom provided for the entire length of any walkway?
☐Yes ☐No   Are standard guardrails provided for walkway surfaces elevated more than 30 inches above any adjacent floor or ground?
☐Yes ☐No   Are bridges provided over conveyors and similar hazards?
☐Yes ☐No   Are trenches properly shored?
☐Yes ☐No   Is all water removed from trenches?

Illumination

☐Yes ☐No   Are exit signs illuminated?
☐Yes ☐No   Does the building have sufficient lighting?
☐Yes ☐No   Are work areas properly illuminated?
☐Yes ☐No   Is lighting intrinsically safe in areas where flammable or combustible vapors may be present?
☐Yes ☐No   Is any lighting so bright that it causes glare?
☐Yes ☐No   Are light bulbs protected from being broken?

Noise

☐Yes ☐No Are noise levels in work areas within acceptable levels?

☐Yes ☐No Are there areas in the workplace where continuous noise levels exceed 85 dBA?

☐Yes ☐No Is an ongoing preventive health program in place to educate employees in safe levels of noise exposures, effects of noise on personal health, and the use of PPE?

☐Yes ☐No Have work areas where noise levels make voice communication difficult been identified and posted?

☐Yes ☐No Are noise levels being measured with a sound level meter or an octave band analyzer and are records being kept?

☐Yes ☐No Have engineering controls been used to reduce excessive noise levels?

☐Yes ☐No If engineering controls are not feasible, are administrative controls like worker rotation being used to minimize employee exposure to excessive noise?

☐Yes ☐No Is approved hearing protective equipment available to every employee working in noisy areas?

☐Yes ☐No Are employees who use ear protectors properly trained and fitted?

☐Yes ☐No Are employees in high noise areas given periodic audiometric testing to ensure an effective hearing protection system is in place?

Sanitation

☐Yes ☐No Are toilet and shower facilities cleaned daily?

☐Yes ☐No Are eating areas kept in a clean and sanitary condition?

☐Yes ☐No Are appliances in eating areas cleaned daily?

☐Yes ☐No Are outside areas free of debris and kept clean and orderly?

☐Yes ☐No Are first-aid kits kept in a sanitary condition?

☐Yes ☐No Are measures taken to prevent the infestation of insects and rodents?

☐Yes ☐No Are toiletries provided in washrooms?

☐Yes ☐No Are containers provided for oily rags?

Exits

☐Yes ☐No Are all exits marked and illuminated by a reliable source of light?

☐Yes ☐No Are all doors or passageways that are not exits appropriately marked?

☐Yes ☐No Is lettering on "Exit" signs at least 5 inches high and 1/2 inch wide?

☐Yes ☐No Are exit doors side-hinged?

☐Yes ☐No Are exits free of obstructions?

☐Yes ☐No Are there enough exits to permit prompt emergency escape?

☐Yes ☐No Where ramps are used, is the slope of any and all ramps limited to 1 vertical foot and 12 horizontal feet?

☐Yes ☐No Are frameless glass doors, glass exit doors, storm doors, etc., fully tempered and do they meet safety requirements for human impact?

☐Yes ☐No Do exit doors open outward away from the direction of exit without requiring the use of a key or any special knowledge or effort?

☐Yes ☐No If panic hardware is installed on an exit door, will it allow the door to open with 15 or fewer pounds of force?

☐Yes ☐No Are exit doors that open onto a street, alley, or parking area provided with adequate barriers and warning to prevent employees from stepping into traffic?

Stairways

☐Yes ☐No Are standard stair rails or handrails installed on all stairways with four or more risers?

☐Yes ☐No Are stairways at least 22 inches wide?

☐Yes ☐No Are stair landing platforms not less than 30 inches in the direction of travel and do they extend 22 inches in width at every 12 feet or less of vertical rise?

☐Yes ☐No Do stairs angle no more than 50 degrees and no less than 30 degrees?

☐Yes ☐No Are stairs of hollow-pan type treads and landings filled to the top edge of the pan with solid material?

☐Yes ☐No Are step risers on stairs uniform from top to bottom?

☐Yes ☐No Do steps have a slip resistant surface?

☐Yes ☐No   Are handrails located 30 to 34 inches above the leading edge of stair treads?
☐Yes ☐No   Is there at least 3 inches of clearance between handrails and the walls they're
           mounted on?
☐Yes ☐No   Where doors or gates open directly onto a stairway, is there a platform provided
           so the swing of the door does not reduce the width of the platform to less than 21
           inches?
☐Yes ☐No   Are handrails capable of withstanding a load of 200 pounds?

Elevated Surfaces
☐Yes ☐No   Are appropriate signs posted to show the load capacity of any elevated surface?
☐Yes ☐No   Are standard guardrails provided on surfaces elevated more than 30 inches
           above the floor or ground?
☐Yes ☐No   Are 4-inch toeboards provided on elevated surfaces that expose personnel or
           machinery to falling objects?
☐Yes ☐No   Is there a permanent means of access and egress to elevated storage and work
           surfaces?
☐Yes ☐No   Is material on elevated surfaces piled, stacked, or racked in a manner that will
           prevent it from tipping, falling, collapsing, rolling, and/or spreading?
☐Yes ☐No   Are dock boards or bridge plates used in transferring materials between docks
           and trucks or rail cars?

Floor and Wall Openings
☐Yes ☐No   Are floor openings guarded by a cover, a guardrail, or the equivalent on all
           sides?
☐Yes ☐No   Are toeboards installed around edges of permanent floor openings?
☐Yes ☐No   Are skylight screens capable of supporting a load of at least 200 pounds?
☐Yes ☐No   Are grates or floor covers used that will not affect foot traffic or rolling equipment?
☐Yes ☐No   Are unused portions of pits that are not in use either covered or protected by
           guardrails or the equivalent?
☐Yes ☐No   Are manhole, trench, or other covers (including their supports) capable of
           carrying a truck rear axle load of at least 20,000 pounds?

EQUIPMENT

Personal Protective Equipment
☐Yes ☐No   Are protective goggles or face shields provided and worn where there is a danger
           of flying particles or corrosive materials?
☐Yes ☐No   Are approved safety glasses worn in areas where risks are present?
☐Yes ☐No   Are protective gloves, aprons, shields, or other means provided where it is
           reasonable to expect that employees may be cut or may be exposed to corrosive
           liquids, chemicals, blood, or other potentially infectious materials?
☐Yes ☐No   Are hard hats provided where the danger of falling objects exists?
☐Yes ☐No   Are hard hats inspected periodically for damage to the shell or suspension
           system?
☐Yes ☐No   Is appropriate foot protection used as required?
☐Yes ☐No   Are approved respirators provided for regular and emergency use as needed?
☐Yes ☐No   Is PPE maintained in a sanitary condition and kept ready for use?
☐Yes ☐No   Are an eye wash and quick drench shower available within a work area where
           employees are exposed to injurious corrosive materials?
☐Yes ☐No   Are adequate work procedures and PPE provided when cleaning up spilled toxic
           or hazardous materials or liquids?
☐Yes ☐No   Are procedures in place for disposing of contaminated PPE?
☐Yes ☐No   Are employees trained in the use, maintenance, limitations, storage, and
           inspection of PPE?
☐Yes ☐No   Where machines and/or equipment are used that process, handle, or apply
           materials that could injure employees, is that machinery or equipment cleaned
           and/or decontaminated before being overhauled or placed in storage?

Ladders

☐Yes ☐No   Are ladders inspected and kept in good condition, free of grease and oil?

☐Yes ☐No   Are ladders and ladder rungs equipped with non-slip feet?

☐Yes ☐No   Are employees trained in the proper use of ladders?

☐Yes ☐No   Is it prohibited to place ladders on unstable bases (such as boxes or barrels) to obtain extra height?

☐Yes ☐No   Are broken or faulty ladders removed from service?

☐Yes ☐No   Are employees instructed to face the ladder when climbing or descending?

☐Yes ☐No   Are employees instructed to not use the top step of stepladders as a step?

☐Yes ☐No   Do ladders in use extend at least 3 feet above the elevated surface?

☐Yes ☐No   Are metal ladders marked with signs cautioning against use around electrical power sources?

☐Yes ☐No   Are rungs of ladders uniformly spaced at 12 inches from the center of one rung to the center of the next?

Compressors and Compressed Air

☐Yes ☐No   Are compressors equipped with pressure relief valves and pressure gauges?

☐Yes ☐No   Are compressor air intakes installed that ensure that only clean, uncontaminated air enters?

☐Yes ☐No   Are air filters inspected regularly?

☐Yes ☐No   Are compressor safety devices checked frequently?

☐Yes ☐No   Before repairs are done on any pressure system, is pressure bled-off and the system locked out?

☐Yes ☐No   Are signs posted warning of automatic starting procedures for compressors?

☐Yes ☐No   Are any belt drive systems totally enclosed?

☐Yes ☐No   Is it strictly prohibited to direct a compressed air flow toward a person?

☐Yes ☐No   Are safety chains or other suitable locking devices used at couplings of high pressure hose lines where a connection failure could create a hazard?

☐Yes ☐No   When compressed air is used with abrasive blast cleaning equipment, is the opening valve a type that must be help open manually?

☐Yes ☐No   Is every compressed air receiver equipped with a pressure gauge and with one or more automatic, spring-loaded safety valves?

☐Yes ☐No   Is the total relieving capacity of the safety valve capable of preventing pressure in the receiver from exceeding the maximum allowable working pressure of the receiver by more than 10 percent?

☐Yes ☐No   Is every air receiver provided with a drain pipe and valve at the lowest point for the removal of accumulated oil and water?

☐Yes ☐No   Are compressed air receivers periodically drained of moisture and oil?

☐Yes ☐No   Are all safety valves tested frequently and regularly to determine whether or not they are in good working condition?

☐Yes ☐No   Is a current operating permit on record?

☐Yes ☐No   Are inlets of air receivers and piping systems free of accumulated oil and carbonaceous material?

Compressed Gas Cylinders

☐Yes ☐No   Are cylinders with a water capacity of over 30 pounds equipped with a means for connecting a valve protector device, or with a collar or recess to protect the valve?

☐Yes ☐No   Are cylinders clearly marked to identify their contents?

☐Yes ☐No   Are compressed gas cylinders stored in areas protected from external heat sources?

☐Yes ☐No   Are cylinders stored or located in areas where they will not be damaged?

☐Yes ☐No   Are cylinders stored or located in areas where they will not be tampered with by unauthorized personnel?

☐Yes ☐No   Are cylinders stored or transported in a way that will prevent them from tipping, falling, or rolling?

☐Yes ☐No   Are valve protector caps placed on cylinders that are not in use or not connected for use?

☐Yes ☐No   Are all valves closed before a cylinder is moved, when it's empty, or at the end of a job?

☐Yes ☐No   Are low-pressure fuel-gas cylinders regularly checked (including a close inspection of the bottom of the cylinder) for defects or wear that might render them unfit for service?

☐Yes ☐No   Are cylinders stored at least 20 feet from highly combustible materials?

☐Yes ☐No   Are bottles maintained with current hydro-inspection dates?

☐Yes ☐No   Are fuel gas and oxygen stored at a minimum of 20 feet apart or separated by a 1-hour fire wall?

☐Yes ☐No   Are in-service cylinders supported to prevent tipping?

Hoist and Auxiliary Equipment

☐Yes ☐No   Is each overhead electric hoist equipped with a limit device to stop the hook travel at its highest and lowest point of safe travel?

☐Yes ☐No   Will each hoist automatically stop and hold any load up to 125% of its rated load if its actuating force is removed?

☐Yes ☐No   Is the rated load of each hoist legibly marked and visible to its operator?

☐Yes ☐No   Are stops provided at the safe limits of travel for hoists?

☐Yes ☐No   Are hoist controls plainly marked to indicate the direction of travel or motion?

☐Yes ☐No   Is each cage-controlled hoist equipped with an effective warning device?

☐Yes ☐No   Are close-fitting guards or other suitable devices installed on hoists to assure hoist ropes will be maintained in the sheave grooves?

☐Yes ☐No   Are hoist chains or ropes sufficient in length to handle the full range of movement of the application while still maintaining two full wraps on the drum at all times?

☐Yes ☐No   Are nip points and contact points between hoist ropes and sheaves which are permanently located within 7 feet of the floor, ground, or working platform properly guarded?

☐Yes ☐No   Is it prohibited to use chains or rope slings that are kinked or twisted?

☐Yes ☐No   Is it prohibited to use the hoist rope or chain wrapped around the load as a substitute for a sling?

☐Yes ☐No   Are operators instructed to avoid carrying loads over people?

☐Yes ☐No   Are hoists and load bearing structures load tested and certified annually?

☐Yes ☐No   Are pelican hooks equipped with a spring loaded safety clip to prevent accidental load release?

Powered Industrial Truck/Forklifts

☐Yes ☐No   Are only trained personnel allowed to use hoists?

☐Yes ☐No   Are only trained personnel allowed to operate industrial trucks?

☐Yes ☐No   Is substantial overhead protective equipment provided on high-lift rider equipment?

☐Yes ☐No   Are required lift-truck operating rules posted and enforced?

☐Yes ☐No   Is directional lighting provided on each industrial truck that operates in an area with less than 2 foot-candles per square foot of general lighting?

☐Yes ☐No   Do industrial trucks have a warning horn or other audible device that can be clearly heard above the normal noise in areas where it is operated?

☐Yes ☐No   Are the brakes on each industrial truck capable of bringing the vehicle to a complete and safe stop when the vehicle is fully loaded?

☐Yes ☐No   Does the parking brake on each industrial truck effectively prevent the vehicle from moving when it is unattended?

☐Yes ☐No   Are industrial trucks operating in areas where flammable gases or vapors, combustible dust, or ignitable fibers may be present in the atmosphere approved for use in such locations?

☐Yes ☐No   Are drive motor shutoff and brakes applied when motorized hand and hand/rider truck control grip is released?

Portable Power Tools and Equipment

☐Yes ☐No   Are grinders, saws, and similar equipment provided with appropriate safety guards?

☐Yes ☐No   Are power tools used with the correct shield, guard, or attachment according to the manufacturer's recommendations?

☐Yes ☐No   Are portable circular saws equipped with guards above and below the base shoe?

☐Yes ☐No   Are circular saw guards checked to ensure that they are in proper condition and that they are not leaving the blade unguarded?

☐Yes ☐No   Are rotating or moving parts of equipment guarded to eliminate the possibility of physical contact?

☐Yes ☐No   Are all cord-connected, electrically operated tools and equipment properly grounded or double insulated?

☐Yes ☐No   Are effective guards in place over belts, pulleys, chains, and sprockets?

☐Yes ☐No   Are portable fans provided with full guards or screens having openings of 1/2 inch or less?

☐Yes ☐No   Is hoisting equipment available for lifting heavy objects?

☐Yes ☐No   Are hoist ratings and characteristics of hoisting equipment appropriate for the task being performed?

☐Yes ☐No   Are ground-fault circuit interrupters provided on all temporary electrical 15-and 20-amp circuits used during periods of construction?

☐Yes ☐No   Are pneumatic or hydraulic hoses on power tools checked regularly for damage and wear?

Hand Tools and Hand-Held Equipment

☐Yes ☐No   Are all tools and equipment kept in good condition?

☐Yes ☐No   Are hand tools that develop mushroomed heads during use (chisels, punches, etc.) reconditioned or replaced as necessary?

☐Yes ☐No   Are broken handles on hammer, axes, and similar equipment replaced promptly?

☐Yes ☐No   Are worn or bent wrenches replaced on a regular basis?

☐Yes ☐No   Are the appropriate handles used on files and similar tools?

☐Yes ☐No   Are employees made aware of the hazards caused by using faulty tools or by using them improperly?

☐Yes ☐No   Are appropriate safety glasses, face shields, etc., used while using tools or equipment that might produce airborne particles or might be subject to breakage?

☐Yes ☐No   Are jacks checked periodically?

☐Yes ☐No   Are tool handles securely attached?

☐Yes ☐No   Are cutting edges on tools kept sharp to prevent skipping or binding?

☐Yes ☐No   Are tools stored in a dry and secure location?

Abrasive Wheel Equipment Grinders

☐Yes ☐No   Is a work rest used and positioned within 1/8 of an inch from the wheel?

☐Yes ☐No   Is an adjustable tongue in place on the top side of the grinder and positioned within 1/4 of an inch from the wheel?

☐Yes ☐No   Do side guards cover the spindle, nut, flange, and 75% of the wheel diameter?

☐Yes ☐No   Are bench and pedestal grinders permanently mounted?

☐Yes ☐No   Are goggles and face shields always worn when grinding?

☐Yes ☐No   Is the maximum RPM rating of each abrasive wheel compatible with the RPM rating of the motor?

☐Yes ☐No   Are fixed or permanently mounted grinders connected to their electrical supply with metallic conduit or other permanent method?

☐Yes ☐No   Does each grinder have its own on/off switch?

☐Yes ☐No   Is each grinder properly grounded?

☐Yes ☐No   Are new abrasive wheels inspected and ring tested before mounting?

☐Yes ☐No   If grinders produce large amounts of dust, are they equipped with dust collectors or powered exhausts?

☐Yes ☐No   Are splash guards installed on grinders that use coolant?

☐Yes ☐No   Is the area around grinders kept clean?

Powder-Actuated Tools

☐Yes ☐No   Do employees who operate powder-actuated tools properly trained?

☐Yes ☐No   Do employees who operate powder-actuated tools carry valid operators cards?

☐Yes ☐No   Are individual powder-actuated tools kept in their own locked container when not in use?

☐Yes ☐No   Are proper signs posted when tools are in use?

☐Yes ☐No   Are powder-actuated tools left unloaded until they are in use?

☐Yes ☐No   Are powder-actuated tools inspected before each use for obstructions or defects?

☐Yes ☐No   Do operators have and use appropriate PPE?

Machine Guarding

☐Yes ☐No   Is a training program in place to train employees in safe machine operation?

☐Yes ☐No   Is adequate supervision provided to ensure that employees are following safe procedures?

☐Yes ☐No   Is there a regular machine safety inspection program?

☐Yes ☐No   Is there evidence that safeguards have been removed, bypassed, or tampered with?

☐Yes ☐No   Is machinery kept clean and in good working condition?

☐Yes ☐No   Is sufficient clearance provided around and between machines to allow for safe operation, servicing, material handling, and waste removal?

☐Yes ☐No   Are equipment and machinery properly secured to prevent tipping or other hazardous movement?

☐Yes ☐No   Is there a power shutoff switch within reach of the operator's position?

☐Yes ☐No   If there are more than one operator are separate controls provided?

☐Yes ☐No   Is there a lockout/tagout program in place?

☐Yes ☐No   Are non-current-carrying metal parts of electrically operated machines bonded or grounded?

☐Yes ☐No   Are foot-operated switches guarded or arranged to prevent accidental actuation?

☐Yes ☐No   Are manually operated control valves and switches clearly identified and readily accessible?

☐Yes ☐No   Are pulleys and belts within 7 feet of the floor or other working surface properly guarded?

☐Yes ☐No   Are all moving chains and gears properly guarded?

☐Yes ☐No   Are splash guards mounted on machines that use coolant?

☐Yes ☐No   Are guards in place that meet at least the minimum OSHA requirements to protect employees from nip points, rotating parts, flying chips, and sparks?

☐Yes ☐No   Are machine guards secure and arranged so that they do not create a hazard?

☐Yes ☐No   If special tools are used for placing or removing material, do they protect the operator's hand(s)?

☐Yes ☐No   Are revolving drums, barrels, or containers guarded by an interlocked device that stops the part from moving unless the guard is in place?

☐Yes ☐No   Do arbors and mandrels have firm and secure bearings, and are they free from play?

☐Yes ☐No   Are machines prevented from starting automatically when power is restored after a power failure or shutdown?

☐Yes ☐No   Are machines free of excessive vibration when the largest size tool is run at full speed?

☐Yes ☐No   If machinery is cleaned with compressed air, is the air pressure controlled and are PPE and other safeguards used to protect operators and other workers from injury.

☐Yes ☐No   Are fan blades protected with a guard with openings no larger than 1/2 inch when they are in operation with 7 feet of the floor?

☐Yes ☐No   Are saws used for ripping equipped with anti-kickback devices and spreaders?

☐Yes ☐No   Are radial arm saws arranged to allow the cutting head to gently return to the back of the table when released?

☐Yes ☐No   Are your electricians familiar with the requirements of the National Electrical Code (NEC)?

☐Yes ☐No   If electrical work is contracted out, do you specify compliance with NEC?

☐Yes ☐No   Are there electrical installations in hazardous dust or vapor areas and do they meet the NEC specifications for hazardous locations?

☐Yes ☐No   Are all electrical cords strung so they do not hang on pipes, nails, hooks, etc.?

☐Yes ☐No   Is all conduit, cable, etc. properly attached to all supports and tightly connected to junction and outlet boxes?

☐Yes ☐No   Is there any evidence of fraying on any electrical cord?

☐Yes ☐No   Are rubber cords kept free of oil, grease, and chemicals?

☐Yes ☐No   Are all metallic cables and conduits properly grounded?

☐Yes ☐No   Are portable electric tools and appliances grounded or double insulated?

☐Yes ☐No   Are all ground connections clean and tight?

☐Yes ☐No   Are fuses and circuit breakers of the right type and size for the load on each circuit?

☐Yes ☐No   Are all fuses free of "jumping" with pennies or metal strips?

☐Yes ☐No   Are switches mounted in tightly closed metal boxes?

☐Yes ☐No   Are all electrical switches marked to show their purpose?

☐Yes ☐No   Are all motors clean and kept free of excessive grease and oil?

☐Yes ☐No   Are all motors properly maintained and provided with adequate overcurrent protection?

☐Yes ☐No   Are bearings in good condition?

☐Yes ☐No   Are all lamps kept free of combustible material?

☐Yes ☐No   Are portable lights equipped with proper guards?

☐Yes ☐No   Is your electrical system checked periodically by someone competent in the NEC?

☐Yes ☐No   Are fusible links in place?

☐Yes ☐No   Are employees required to report as soon as possible any obvious hazard to life or property resulting from electrical equipment or lines?

☐Yes ☐No   Are employees trained to make preliminary inspections and/or tests to determine what conditions exist before starting work on electrical equipment or lines?

☐Yes ☐No   When electrical equipment or lines are serviced, are necessary switches opened, locked out, and tagged whenever possible?

☐Yes ☐No   Do extension cords have a grounding conductor?

☐Yes ☐No   Are multiple plug adapters prohibited?

☐Yes ☐No   Are ground-fault circuit interrupters installed on each temporary 15-or 20-ampere, 120 volt AC circuit at locations where construction, demolition, modifications, alterations, or excavations are being performed?

☐Yes ☐No   Are temporary circuits protected by suitable disconnecting switches or plug connectors at the junction with permanent wiring?

☐Yes ☐No   Are flexible cords and cables free of splices or taps?

☐Yes ☐No   Are cord, cable, and raceway connections intact and secure?

☐Yes ☐No   Is the location of electrical power lines and cables (overheads, underground, under floor, other side of walls, etc.) determined before digging, drilling, or similar work is done?

☐Yes ☐No   Is the use of metal ladders prohibited in areas where the ladder or the user could come in contact with energized parts of equipment, fixtures, or conductors?

☐Yes ☐No   Are disconnecting means always opened before fuses are replaced?

☐Yes ☐No   Is sufficient access and working space provided and maintained around all electrical equipment to permit safe operations and maintenance?

☐Yes ☐No   Are unused openings (including conduit knockouts) in electrical enclosures and fittings protected with appropriate covers, plugs, or plates?

☐Yes ☐No   Are disconnecting switches for electrical motors in excess of 2 horsepower, capable of opening the circuit when the motor is in a stalled condition, without exploding? Note that switches must be rated equal to or in excess of the motor hp rating.

☐Yes ☐No   Is each motor disconnecting switch or circuit breaker located within sight of the motor control device?

☐Yes ☐No   Is each motor located within sight of its controller or the controller disconnecting means capable of being locked in the open position, or is a separate disconnecting means installed in the circuit within sight of the motor?

☐Yes ☐No   Is the controller for each motor in excess of two horsepower, rated in horsepower equal to or in excess of the rating of the motor it serves?

☐Yes ☐No   Are employees who regularly work on or around energized electrical equipment or lines instructed in CPR?

☐Yes ☐No   Are employees prohibited from working alone on energized lines or on equipment over 600 volts?

WELDING, CUTTING, AND BRAZING

☐Yes ☐No   Are only authorized and trained individuals permitted to use welding, cutting, and/or brazing equipment?

☐Yes ☐No   Does each operator have a copy of and follow appropriate operating instructions?

☐Yes ☐No   Are compressed gas cylinders examined regularly for defects or signs of deep rusting and/or leakage?

☐Yes ☐No   Are precautions taken to prevent the mixture of oxygen and flammable gases other than at a burner or in a standard torch?

☐Yes ☐No   Are only approved apparatus used (torches, regulators, pressure-reducing valves, acetylene generators, manifolds)?

☐Yes ☐No   Are cylinders kept away from heat sources?

☐Yes ☐No   Are cylinders kept away from elevators, stairs, or gangways?

☐Yes ☐No   Are hot work permits required?

☐Yes ☐No   Are used drums, barrels, tanks, and other containers thoroughly cleaned so that no explosive or hazardous chemical substances remain?

☐Yes ☐No   Is required PPE used properly and inspected on a regular basis?

☐Yes ☐No   Has an inspection been made to ensure adequate ventilation where welding or cutting is conducted?

☐Yes ☐No   Are environmental monitoring tests performed when work occurs in confined spaces, and are means provided from quick emergency egress?

LOCKOUT/TAGOUT PROCEDURES

☐Yes ☐No   Is it mandatory that all hazardous energy sources be de-energized, disengaged, blocked, or locked-out during cleaning, servicing, adjusting, or setting-up operations?

☐Yes ☐No   When electrical control circuits cannot be disconnected, are the appropriate electrical enclosures identified?

☐Yes ☐No   When electrical control circuits cannot be disconnected, is a means provided to assure that the control circuit can also be disconnected and locked-out?

☐Yes ☐No   Is the lockout of control circuits in lieu of locking out main power disconnects prohibited?

☐Yes ☐No   Are all equipment control valve handles provided with a means for lockout?

☐Yes ☐No   Do standard lockout procedures require that stored energy be released or blocked?

☐Yes ☐No   Are appropriate employees provided with individually keyed safety locks?

☐Yes ☐No   Are employees required to keep personal control of their key(s) while their safety locks are in use?

☐Yes ☐No   Is it required that only those employees who are exposed to a particular hazard be the only ones to place or remove the safety lock?

☐Yes ☐No   Is it required that employees verify equipment lockout by attempting a start-up after making sure no one is exposed?

☐Yes ☐No   Are employees instructed to always push the control circuit stop button prior to re-energizing the main power switch?

☐Yes ☐No   Is it possible to identify any or all employees working on locked-out equipment by their locks or accompanying tags?

☐Yes ☐No   If equipment lines cannot be shut down, locked, or tagged out, is a safe procedure established and rigidly followed?

CONFINED SPACES

☐Yes ☐No   Are confined spaces thoroughly emptied of corrosive or hazardous substances before entry?

☐Yes ☐No   Are all lines to a confined space containing hazardous substances locked and tagged out before entry?

☐Yes ☐No   Is adequate ventilation provided prior to a confined space entry?

☐Yes ☐No   Are appropriate atmospheric tests performed prior to confined space entry?

☐Yes ☐No   Is the atmosphere inside the confined space frequently tested or continuously monitored during work?

☐Yes ☐No   Are atmospheric tests done at all levels, from top to bottom?

☐Yes ☐No   Is adequate illumination provided in confined spaces?

☐Yes ☐No   Is a safety observer assigned outside of the confined space?

☐Yes ☐No   Is the safety observer appropriately trained and equipped to handle emergencies?

☐Yes ☐No   Is approved respiratory equipment required if the atmosphere inside the confined space cannot be made acceptable?

☐Yes ☐No   Is portable electrical equipment used inside confined spaces either grounded or insulated or equipped with ground fault protection?

☐Yes ☐No   Are hot work permits required before welding and other oxygen-consuming equipment is used?

☐Yes ☐No   Does the safety observer have the authorization to shut down a job if needed?

FLAMMABLE AND COMBUSTIBLE MATERIALS

☐Yes ☐No   Are combustible materials stored in covered metal receptacles and removed from work areas promptly?

☐Yes ☐No   Are approved containers and tanks used for the storage and handling of flammable and combustible liquids?

☐Yes ☐No   Are flammable liquids kept in closed containers when not in use?

☐Yes ☐No   Are bulk drums of flammable liquids grounded and bonded to containers during dispensing?

☐Yes ☐No   Do storage rooms have adequate ventilation and explosion-proof lights?

☐Yes ☐No   Are no smoking signs posted on liquefied petroleum gas tanks and in areas where flammable or combustible materials are used and stored?

☐Yes ☐No   Are fuel gas cylinders separated by distance or fire-resistant barriers, etc., while in storage?

☐Yes ☐No   Are spills of flammable and combustible liquids promptly cleaned up?

☐Yes ☐No   Are storage tanks adequately vented to prevent excessive vacuum or pressure?

FUELING OPERATIONS

☐Yes ☐No   Is it prohibited to conduct fueling operations while the engine is running?

☐Yes ☐No   Are fuel tank caps replaced and secured before the engine is started?

☐Yes ☐No   Is there always metal contact between the container and the fuel tank?

☐Yes ☐No   Are fueling hoses designed to handle the specific type of fuel being dispensed?

☐Yes ☐No   Are fueling operations prohibited in buildings or other enclosed areas that are not specifically vented for that purpose?

☐Yes ☐No   Are nozzles of the self-closing type where fueling or fuel transfer is done through a gravity flow system?

HAZARDOUS CHEMICALS

☐Yes ☐No   Are employees trained in the safe use of hazardous chemicals and materials?

☐Yes ☐No   Are employees knowledgeable of potential workplace chemical hazards?

☐Yes ☐No   Are eye wash fountains and safety showers provided in areas where corrosive chemicals are handled?

☐Yes ☐No   Are containers labeled?

☐Yes ☐No   Are employees required to use personal protective clothing and equipment when handling chemicals?

☐Yes ☐No   Are flammable and toxic chemicals kept in closed containers when not in use?

☐Yes ☐No   Are chemical piping systems clearly marked to identify their content?

☐Yes ☐No   Are adequate means readily available for containing spills or overflows properly and safely?

☐Yes ☐No   Have standard operating procedures been established and are they being followed when cleaning up chemical spills?

☐Yes ☐No   Are respirators stored in a convenient, clean, and sanitary location?

☐Yes ☐No   Are the respirators intended for emergency use adequate for the specific uses for which they are intended?

☐Yes ☐No   Are employees prohibited from eating in areas where hazardous chemicals are present?

☐Yes ☐No   Is PPE provided, used, and maintained where it is needed?

☐Yes ☐No   Are written standard operating procedures in place for the selection and use of respirators in situations where respirators are necessary?

☐Yes ☐No   Are employees who use respirators instructed/trained in the correct use and limitations of the respirators?

☐Yes ☐No   Are respirators regularly inspected, cleaned, sanitized, and maintained?

☐Yes ☐No   Do you have a medical or biological monitoring system in place where hazardous substances are used in your processes?

☐Yes ☐No   Are control procedures regarding respirators, ventilation, handling practices, etc., instituted for hazardous materials where appropriate?

☐Yes ☐No   Are hazardous substances handled in properly designed and exhausted booth locations?

☐Yes ☐No   Is carbon monoxide kept within acceptable levels where internal combustion engines are used?

☐Yes ☐No   Is vacuuming used for clean up whenever possible instead of blowing or sweeping?

☐Yes ☐No   Are materials that give off toxic asphyxiant, suffocating, or anesthetic fumes stored in remote or isolated locations when not in use?

☐Yes ☐No   Are annual spirometry and medical examinations maintained for personnel using respirators?

☐Yes ☐No   Are areas that use cryogenic nitrogen or carbon dioxide equipped with oxygen level monitors and warning devices?

HAZARDOUS SUBSTANCES COMMUNICATION

☐Yes ☐No   Is a list of the hazardous substances maintained in your workplace maintained?

☐Yes ☐No   Is there a current written exposure control plan in place for occupational exposure to bloodborne pathogens and other potentially infectious materials?

☐Yes ☐No   Is there a written hazard communication program in place that deals with Material Safety Data Sheets (MSDS), labeling, and employee training?

☐Yes ☐No   Are containers for hazardous substances (i.e., vats, bottles, storage tanks, etc.) labeled to identify the products they contain and to warn of any potential hazard (communication of specific health or physical hazards)?

☐Yes ☐No   Are MSDSs readily available for each hazardous substance used?

☐Yes ☐No   Are employees trained to 1.) recognize tasks that might result in occupational exposure; 2.) use work practice, engineering controls, and PPE and know their limitations; 3.) obtain information on the types, selection, use, location, removal, handling, decontamination, and disposal of PPE; and 4.) carry out an emergency response plan?

IDENTIFICATION OF PIPING SYSTEMS

☐Yes ☐No   When non-potable water is piped through a facility, are outlets or taps posted to alert employees that it is unsafe and not be used for drinking or other personal use?

☐Yes ☐No   When hazardous substances are transported through above ground piping, is each pipeline identified at points where confusion could introduce hazards to employees?

☐Yes ☐No   When pipelines are identified by color, are all visible parts of the line so identified?

☐Yes ☐No   When pipelines are identified by color bands or tapes, are the bands or tapes placed at reasonable intervals and at each outlet, valve, or connection?

☐Yes ☐No   When pipelines are identified by color, is the color code posted at all locations where confusion could introduce hazards to employees?

☐Yes ☐No   When the contents of pipelines are identified by name or abbreviation, is the information readily visible on the pipe near each valve or outlet?

☐Yes ☐No   When pipelines are identified by tags, are the tags constructed of durable materials, are they clearly and permanently distinguishable, and are they installed at each valve or outlet?

☐Yes ☐No   When pipelines are heated by electricity, steam, or other external source, are suitable warning signs or tags placed at unions, valves, or other serviceable parts of the system?

MATERIAL HANDLING

☐Yes ☐No   Is there safe aisle and doorway clearance for equipment?

☐Yes ☐No   Are aisles properly marked and cleared?

☐Yes ☐No   Are motorized vehicles and mechanized equipment inspected daily or at least prior to use?

☐Yes ☐No   Are vehicles shut off and braked prior to loading or unloading?

☐Yes ☐No   Are containers of combustibles or flammable materials always separated by dunnage sufficient to provide stability when being stacked or while being moved?

☐Yes ☐No   Are dock boards (bridge plates) used when loading or unloading operations are taking place between vehicles and docks?

☐Yes ☐No   Are trucks and trailers secured from movement during loading and unloading operations?

☐Yes ☐No   Are dock plates and loading ramps constructed and maintained with sufficient strength to support loading?

☐Yes ☐No   Are hand trucks maintained to be in safe operation condition?

☐Yes ☐No   Are chutes equipped with sideboards of sufficient height to prevent the materials being handled from falling off?

☐Yes ☐No   Are chutes and gravity roller sections firmly placed or secured to prevent displacement?

☐Yes ☐No   At the delivery end of the rollers or chutes, are provisions made to brake the movement of the handled materials?

☐Yes ☐No   Are pallets inspected before being loaded or moved?

☐Yes ☐No   Are hooks with safety latches or other arrangements used when hoisting materials so that slings or load attachments won't slip off the hoist hooks?

☐Yes ☐No   Are securing chains, ropes, chokers, or slings adequate for the job they perform?

☐Yes ☐No   When hoisting material or equipment, are provisions made to assume no one will be passing under suspended loads?

☐Yes ☐No   Is material interlaced or stacked to prevent sliding or tipping?

☐Yes ☐No   Are shelves secure and constructed to withstand maximum storage weight?

☐Yes ☐No   Are shelves secured to prevent tipping or falling?

TRANSPORTATION OF EMPLOYEES AND MATERIALS

☐Yes ☐No   Do employees who operate vehicles on public roads have valid licenses?

☐Yes ☐No   Do employees who operate vehicles (vans, buses, or trucks, etc.) carrying seven or more employees have the appropriate license?

☐Yes ☐No   Are vans, buses, or trucks used to transport employees equipped with adequate seats and seat belts?

☐Yes ☐No   When trucks are used to transport employees, are adequate provisions made to prevent them from falling from such vehicles?

☐Yes ☐No   Are vehicles used to transport employees equipped with lamps, brakes, horns, mirrors, windshields, and turn signals in good working order?

☐Yes ☐No   Are transport vehicles provided with handrails, steps, stirrups, or similar devices and are those devices placed to allow employees to safely mount and dismount the vehicles?

☐Yes ☐No   Are employee transport vehicles equipped with at least two reflective type flares?

☐Yes ☐No   Is each employee transport vehicle equipped with a fully charged fire extinguisher in good condition with at least a 4 B:C rating?

☐Yes ☐No   When cutting tools or tools with sharp edges are carried in passenger compartments, are they placed in closed containers that are secured?

☐Yes ☐No   Are employees prohibited from riding on top of any load that can shift, topple, or become unstable?

SPRAYING OPERATIONS

☐Yes ☐No   Is adequate ventilation assured before spray operations are started?

☐Yes ☐No   Is mechanical ventilation provided when spraying operations are done within enclosed areas?

☐Yes ☐No   Does mechanical venting apparatus properly vent contaminated air?

☐Yes ☐No   Are spraying areas free of hot surfaces?

☐Yes ☐No   Are spraying areas at least 20 feet from flames, sparks, operating electrical motors, or other ignition sources?

☐Yes ☐No   If portable lamps are used to illuminate spraying areas, are they suitable for use in a hazardous area?

☐Yes ☐No   If appropriate, is approved respiratory equipment provided and used during spraying operations?

☐Yes ☐No   Do solvents used for cleaning have a flash point of 100 degrees or more?

☐Yes ☐No   Are fire control sprinkler heads kept clean?

☐Yes ☐No   Are "No Smoking" signs posted in spray areas, paint rooms, paint booths, and paint storage areas?

☐Yes ☐No   Are spraying areas kept clean of combustible residue?

☐Yes ☐No   Are spray booths constructed of metal, masonry, or other substantial noncombustible material?

☐Yes ☐No   Are spray booth floors and baffles noncombustible and easily cleaned?

☐Yes ☐No   Is infrared drying apparatus kept out of the spray area during spraying operations?

☐Yes ☐No   Are spraying booths completely ventilated before drying apparatus is used?

☐Yes ☐No   Is electric drying apparatus properly grounded?

Glossary

Abatement - elimination of hazards.

Abrasions - scrapes, cuts, or any damage resulting from friction.

Accessible - having access; right of access.

Acids - having a pH of less than seven; can cause damage to materials; caustic.

Affected person - (for lockout/tagout) a person affected by an activity, such as an operator of a piece of equipment that has been locked or tagged out of service.

ANSI - American National Standards Institute; sets standards for industry.

Approved - In compliance with, or listed by, a nationally recognized testing laboratory

Attendant - person who is trained to observe and control confined space entrants without entering the space.

Authorized person - a person authorized by the employer to perform a specific activity (e.g., lockout/tagout, confined entry)

Baseline test - a test designed to give a starting reference point for future exposure to hazardous materials or conditions.

Biannual - occurs every 6 months.

Blanking or blinding - the absolute closure of a pipe, line, or duct by placing a solid plate (skillet) between two flanges to cover the bore in order to prevent flow. It is important to make sure the skillet can withstand the pressure.

Blood - human blood, human blood components, and products made from human blood.

Bloodborne - contained in the blood.

Bloodborne pathogens - pathogenic microorganisms that are present in human blood and can cause disease in humans. These pathogens include, but are not limited to, hepatitis B virus (HBV) and human immunodeficiency virus (HIV).

Carcinogen - a general term meaning an agent that causes cancer.

Catastrophic release - major uncontrolled emission, fire, or explosion involving one or more highly hazardous chemical that could harm employees.

Certified - a term to describe equipment that has been tested and approved by a national testing agency; a person who has a certification.

Chock - to block forward or backward movement of wheeled vehicles.

Circuit - a system that contains the flow of electricity.

Class A Fire - fire resulting form the ignition of ordinary combustibles such as wood or paper; an ash will result.

Class B Fire - fire resulting from the ignition of flammable liquids.

Class C Fire - fire resulting from the ignition of energized electric circuitry.

Class D Fire - fire resulting from the ignition of metal dust, powder, or particulate.

Closed Container - A container sealed by means of a lid or other device that neither liquid nor vapor will escape from it at ordinary temperatures.

Combustible - as a noun, any material that has the composition to add the fuel portion of the chain of fire (e.g. wood, paper) or may be used as an adjective *(see Combustible Liquid)*

Combustible Liquid - OSHA definition - A liquid with a flash point at or above 100° F, but below 200° F. DOT definition - A liquid with a flash point at or above 141° F, but below 200° F.

Competent person - any person who has experience and training on a given topic.

Compressed Gas - a gas or mixture of gases having, in a container, an absolute pressure exceeding 40 psi at 70° degrees F or an absolute pressure exceeding 104 psi at 130° F regardless of the pressure at 70° F.

Conduction - heat transfer through direct contact.

Confined space - a space that is large enough to enter, has limited openings or exits, and is not designed for continuous human occupancy.

Contaminated - the presence or the reasonably anticipated presence of blood or other potentially infectious materials on an item or surface.

Convection - heat transfer through air currents.

Corrosive - A chemical that causes visible destruction of, or irreversible alterations in, living tissues by chemical action at the site of contact.

Coupling device - device used to join objects.

CSP - Certified Safety Professional

Cumulative trauma disorder - injury to tissue, tendons, nerves form repeated stress or strain.

D

dBA - noise (decibels) measured on the A-weighted scale to determine noise exposure limits.

Dead man switch - switch that must be engaged simultaneously with the primary switch to allow operation.

Decibels - unit for measuring the relative loudness of sounds.

Decontaminate - to remove all infectious or contaminating properties.

Decontamination - the use of physical or chemical means to remove, inactivate, or destroy bloodborne pathogens on a surface or item to the point where they are no longer capable of transmitting infectious particles and the surface or item is rendered safe for handling, use, or disposal.

Deficiencies - not complete.

Disconnect switch - switch that must be engaged to allow completion of the electrical circuit.

Disinfected - sterile; will not cause infection.

E

Energy source - an energy source may be electrical, hydraulic, thermal, or pneumatic. Also included is the energy stored in springs and electrical capacitors and the potential energy from suspended parts.

Engineering controls - controls that are designed through engineering concepts and principles to prevent or minimize hazards in the workplace.

Entry permit - the written document that authorizes work to be performed in a permit-required confined space. Entry permits expire at the end of one shift or whenever a space has been vacated for more than 30 minutes.

Entry supervisor - the trained individual who is responsible for coordination of all work activities in permit required confined spaces.

Exit access - the portion of an exit route that leads to an exit. An example of an exit access is a corridor on the fifth floor of an office building that leads to a 2-hour fire resistance-rated enclosed stairway (the Exit).

Exit discharge - the part of the exit route that leads directly outside or to a street, walkway, refuge area, public way, or open space with access to the outside. An example of an exit discharge is a door at the bottom of a 2-hour fire resistance-rated enclosed stairway that discharges to a place of safety outside the building.

Exit - that portion of an exit route that is generally separated from other areas to provide a protected way of travel to the exit discharge. An example of an exit is a 2-hour fire resistance-rated enclosed stairway that leads from the fifth floor of an office building to the outside of the building.

Exit route - a continuous and unobstructed path of exit travel from any point within a workplace to a place of safety (including refuge areas). An exit route consists of three parts: The exit access; the exit; and, the exit discharge. (An exit route includes all vertical and horizontal areas along the route.)

Exposure incident - a specific eye, mouth, other mucous membrane, non-intact skin, or parenteral contact with blood or other potentially infectious materials that result from the performance of an employee's duties.

F

Fastening devices - devices used to fasten or connect objects.

Fiber core - cable or sling core containing natural or synthetic fibers.

Fire area - an area of a building separated from the remainder of the building by construction having a fire resistance of at least 1 hour and having all communicating openings (doors, etc.) properly protected by an assembly having a fire resistance rating of at least 1 hour.

Fire brigade - private or internal fire fighting team that is employed by private sources for the primary protection of that facility.

Fire detection systems - any system that has a primary function of detecting fire.

First aid - any one-time treatment for an injury, and any follow-up visit for the purpose of observation.

Fixed containers - any container that is designed for use in fixed locations.

Flammable liquid - a liquid with a flash point below 100° F.

Flashpoint - minimum temperature at which a liquid gives off enough vapor in sufficient concentration to form an ignitable mixture with air near the surface of the liquid.

Floor hole - an opening measuring less than 12 inches but more than 1 inch in its least dimension, in any floor, platform, pavement, or yard, through which materials but not persons may fall; such as a belt hole, pipe opening, or slot opening.

Floor opening - an opening measuring 12 inches or more in its least dimension, in any floor, platform, pavement, or yard through which persons may fall; such as a hatchway, stair or ladder opening, pit, or large manhole. Floor openings occupied by elevators, dumb waiters, conveyors, machinery, or containers are excluded from this subpart.

Ground fault circuit interrupters - protective device used to interrupt electrical current in the event of accidental ground.

Hand washing facilities - a facility providing an adequate supply of running potable water, soap, and single use towels or hot-air drying machines.

Handrail - A single bar or pipe supported on brackets from a wall or partition, as on a stairway or ramp, to furnish persons with a handhold in case of tripping.

HazComm - hazard communication.

HazMat - hazardous materials.

HBV - hepatitis B virus.

High hazard area - an area inside a workplace in which operations include high hazard materials, processes, or contents.

HIV - human immunodeficiency virus.

HMIS - refers to the Hazardous Materials Identification System. This system is used by AIH to identify hazards associated with chemicals in the workplace that are not already labeled with hazard warning data. Corresponding codes may be found in the Hazard Index subsection.

Hot work - any work that involves high temperatures (e.g., welding, cutting, brazing.)

Hot work permit - the employer's written authorization to perform operations such as riveting, welding, cutting, burning, and heating, i.e., operations that are capable of providing a source of ignition.

Housekeeping - the act of keeping work areas neat, clean, and orderly.

Ignition - the act of igniting; to initiate a fire or explosion.

Immunizations - shots or injections used in the prevention of disease.

Impregnated - contained within; to be filled with.

Infectious materials - materials that have the capacity to cause infection.

Ingoing nip-points - the point of operation at which two circular devices, rotating in opposite directions, converge to form a nip or a pinch.

Ionizing radiation - radiation that produces electromagnetic waves or particles capable of producing ions.

LEL (lower explosive limit) - the minimum concentration of a substance that will burn or explode in the presence of an ignition source.

Licensed health care professional - a person whose legally permitted scope of practice allows him or her to independently perform the activities required by the Standard for Hepatitis B Vaccination and Post-exposure Evaluation and Follow-up.

Load capacity - the maximum load that can be applied to a structure without failure.

Load rating - the recommended maximum load that cannot cause structure failure.

Lockout (LO) - a term that refers to a system of procedures and training used to notify and remind employees to trace and disengage all energy sources that could cause machinery/equipment to cycle or start during servicing and/or maintenance. An energy source may be electrical, hydraulic, thermal, or pneumatic. Also included is the energy stored in springs and electrical capacitors and the potential energy from suspended parts.

Lockout device - a device that utilizes a positive means such as a lock to hold an energy-isolating device in the safe position and prevents the energizing of a machine or equipment. Locks must be standardized and easily identifiable as lockout devices either by color, shape, or size. Locks must be substantial in order to prevent removal, except by excessive force from special tools such as bolt cutters or other metal cutting tools. Locks must clearly identify the employee who applies them.

Lost work day - any day after the day of injury that the employee would have worked if an occupational illness or injury had not occurred.

Material Safety Data Sheets (MSDS) - a chemical-specific document that contains information on that chemical.

Menace - to threaten; to cause fear or harm.

Metal mesh - containing multiple metal wires interlocked to form a lifting device.

Methodologies - methods derived from extensive study.

MSDS - Material Safety Data Sheet(s), data sheets that contain safety information about chemical products.

Musculoskeletal disorder - physical disorders associated with strains, repetitive motions, general stress to the human body that affect muscles and bones.

Natural - derived from nature; not man-made.

Nonfiber core - core containing metal fibers.

Nonpotable water - water not intended for consumption; not proven safe for consumption.

Occupant load - the total number of persons who may occupy a workplace or portion of a workplace at any one time. The occupant load of a workplace is calculated by dividing the gross floor area of the workplace or portion of a workplace by the occupant load factor for that particular type of workplace occupancy. Information regarding "Occupant load" is located in NFPA 101-2000, Life Safety Code.

Occupational exposure - reasonably anticipated skin, eye, mucous membrane, or parenteral contact with blood or other potentially infectious materials that may result from the performance of an employee's duties.

Occupational hazard - any hazard associated with the performance of labor for a wage or compensation.

OSHA 300 Log - a governmental document provided by the Department of Labor (OSHA) on which "recordable" injuries and illnesses are annotated and posted for employees to view.

Other potentially infectious materials - (1) the following human body fluids: semen, vaginal secretions, cerebrospinal fluid, synovial fluid, pleural fluid, pericardial fluid, peritoneal fluid, amniotic fluid, saliva in dental procedures, any body fluid that is visibly contaminated with blood, and all body fluids in situations where it is difficult or impossible to differentiate between body fluids; (2) Any unfixed tissue or organ (other than intact skin) from a human (living or dead); and (3) HIV-containing cell or tissue cultures, organ cultures, and HIV- or HBV-containing culture medium or other solutions; and blood, organs, or other tissues from experimental animals infected with HIV or HBV.

Oxygen deficient - atmosphere in which there is less than 19.5% oxygen by volume.

Parenteral - piercing mucous membranes or the skin barrier through such events as needle sticks, human bites, cuts, and abrasions.

Pathogens - organisms that can cause harm or disease.

Periodic - occurring at defined time intervals.

Personal protective equipment (PPE) - specialized clothing or equipment (safety shoes, glasses, aprons, etc.) worn by an employee for protection against a hazard. General work clothes (e.g., uniforms, pants, shirts, or blouses), not intended to function as protection against a hazard, are not considered to be personal protective equipment.

Pipetting - the act of "sucking" a liquid through a cylinder, e.g., straws or tubes.

Platform - a working space for persons, elevated above the surrounding floor or ground; such as a balcony or platform for the operation of machinery and equipment.

Pneumatic tool - any tool powered by pressured air.

Point of operation - the precise point at which the material and the tool or machine come into contact.

Portable containers - containers used to transfer contents from one place to another, e.g. gasoline cans.

Portable tank - a closed container having a liquid capacity over 60 gallons (US) and not intended for fixed installation

Potable water - water that is safe for human consumption.

Process - activities involving highly hazardous chemicals including use, storage, manufacturing, movement, or handling.

Process hazard analysis - a systematic method of analyzing any process to determine its hazards.

Professional engineer (PE) - any engineer who has successfully passed the certification exam by the appropriate agency for that state.

Putrid - having an offending odor.

Qualified person - any person who has undergone adequate (certificate) training on a given topic.

Recordable illness or injury - any illness or injury that results in a fatality, lost workday(s), transfer or termination of employment, requires medical treatment other than first aid, involves loss of consciousness, or involves loss of motion.

Refuge area - either:

- o A floor with at least two spaces, separated from each other by smoke-resistant partitions, in a building protected throughout by an automatic sprinkler system that complies with §1910.159.

- o A space along an exit route protected from the effects of fire by separation from other spaces within the building by a barrier with at least a 1-hour fire resistance-rating; or

Regulated waste - liquid or semi-liquid blood or other potentially infectious materials; contaminated items that would release blood or other potentially infectious materials in a liquid or semi-liquid state if compressed; items that are caked with dried blood or other potentially infectious materials and are capable of releasing these materials during handling; contaminated sharps; and pathological and microbiological wastes containing blood or other potentially infectious materials.

Retention - the period for retaining; length of time to be kept.

Runway - a passageway for persons, elevated above the surrounding floor or ground level, such as a foot-walk along shafting or a walkway between buildings.

S

Safety can - an approved container of not more than 5 gallons capacity, having a spring-closing lid and spout cover, and so designed that it will safely relieve internal pressure when subjected to fire exposure. Modification to such a container is not allowed.

Safety cans - cans and containers built with specific safety devices.

Satellite waste container - a drum or other container (up to 55 gallons) that stores hazardous waste at or near the point of generation. Once the container is full, it is considered an accumulated waste.

Self-luminous - a light source that is illuminated by a self-contained power source (e.g., tritium) and that operates independently from external power sources. Batteries are not acceptable self-contained power sources. The light source is typically contained inside the device.

Sharps - medical objects capable of penetrating the skin, (e.g., needles).

Shored - in earth excavation, a method of preventing cave-ins by using materials that will block falling earth.

Sloped - in earth excavation, a method of preventing cave-ins by sloping the top of the excavation away from the bottom.

Solidity - the measure of how solid something is.

Stable - non-reactive; not prone to move easily; reliable.

Stair railing - a vertical barrier erected along exposed sides of a stairway to prevent falls of persons.

Standard railing - a vertical barrier erected along exposed edges of a floor opening, wall opening, ramp, platform, or runway to prevent falls of persons.

Standard strength and construction - any construction of railings, covers, or other guards that meets the requirements of §1910.23 Guarding Floor and Wall Openings and Holes.

Supervisor – for the purposes of this text, supervisor is interchangeable with the terms director, manager, lead, and/or safety representative.

Synthetic - man-made; not normally found in nature.

T

Tag(s) - a prominent warning device and a means of attachment, which is securely fastened to an energy-isolating device (lockout device, or lock) in accordance with an established procedure, to indicate the name of the person applying the tag and that the energy-isolating device (LO device) shall not be removed and the equipment being controlled may not be operated. Tags may only be removed by the person identified on the tag or by his/her supervisor acting under a tag removal procedure. Tags shall never be bypassed, ignored, or otherwise defeated. Tags shall be used with each energy-securing lock/device installed.

Examples:

DO NOT CLOSE	DO NOT OPERATE
DO NOT START	DO NOT OPEN

Tag-line - a rope or line designed to control the movement of suspended loads.

Toeboard - a vertical barrier at floor level erected along exposed edges of a floor opening, wall opening, platform, runway, or ramp to prevent falls of materials.

TWA (time-weighted average) - the computed average measurement of an exposure (toxic, noise, etc.) to a worker, usually over an 8-hour period.

UEL (upper explosive limit) - the highest concentration of a substance that will burn or explode in the presence of an ignition source.

Universal precautions - an approach to infection control. According to the concept of Universal Precautions, all human blood and certain human body fluids are treated as if known to be infectious for HIV, HBV, and other bloodborne pathogens.

UPK - universal precaution kit.

Ventilation - as specified in this section, ventilation involves the prevention of fire and explosion. It is considered adequate to prevent accumulation of significant quantities of vapor-air mixtures in concentration of over 1/4 of the lower flammable limit.

Wall hole - an opening less than 30 inches but more than 1 inch high, of unrestricted width, in any wall or partition; such as a ventilation hole or drainage scupper.

Wall opening - an opening at least 30 inches high and 18 inches wide, in any wall or partition, through which persons may fall; such as a yard-arm doorway or chute opening.

Washing facility - any facility with the primary purpose of being used for cleaning.

Waste-products - products that are not reusable or that can be used.

Water closet - an enclosed toilet for the use of defecation or urination.

Wet process - any process that can result in the work area becoming wet.

Work practice controls - controls that reduce the likelihood of exposure by altering the manner in which a task is performed (e.g., prohibiting recapping of needles by a two-handed technique).

INDEX